Maintaining
and
Troubleshooting
Electrical
Equipment

Maintaining and Troubleshooting Electrical Equipment

Roy Parks
and
Terry Wireman

INDUSTRIAL PRESS INC.

Library of Congress Cataloging-in-Publication Data

Parks, Roy.
  Maintaining and troubleshooting electrical equipment.

  Includes index.
  1. Electric apparatus and appliances—Maintenance and repair.
I. Wireman, Terry.   II. Title.
TK452.P375   1987   621.31′042   87–4234
ISBN 0–8311–1164–X

INDUSTRIAL PRESS INC.
200 Madison Avenue
New York, New York 10016

  6 8 9 7 5

Maintaining and Troubleshooting Electrical Equipment

Composition and art by Tec Set Limited, Surrey, England.

Preface

THE PURPOSE OF THIS BOOK is to present the technical subject of maintaining and troubleshooting electrical equipment in as nontechnical language as possible. While certain terminology must be used, care is taken to avoid unnecessary use of language not familiar to a beginning student and to develop the meaning of new terms as they are introduced. The book is designed for use in industrial training for apprentices and in refresher training for journeymen. Vocational and trade schools should also find the book useful.

The use of advanced mathematics has been avoided and only very basic algebra is required for complete understanding of the material presented. Many well-developed textbooks are available for mathematical analysis if further study is desired.

The beginning student in industrial electricity should study the material in the order in which it is presented and each section should be thoroughly understood before proceeding. The sections on dc instruments and control devices should be supplemented with literature published by manufacturers of control components in the student's work area. This will not only broaden the student's understanding of various applications, but will give specific information concerning applications in the student's work area.

Roy Parks
Terry Wireman

Contents

Maintaining and Troubleshooting Electrical Equipment

Chapter 1 Basic Electricity

ELECTRICAL CIRCUITS, no matter how complex, follow certain basic principles. Knowledge of these principles is essential to understand how an electrical circuit operates and to troubleshoot effectively. An electrical circuit is an arrangement or configuration of components that form a closed loop. There are three parameters in all electrical circuits: current, voltage, and resistance.

Current

Current is a measure of the electron flow in a circuit. This may be compared to the flow of fluid through a pipe, which is measured in gallons per minute (gpm). The current flow of electrons in an electrical circuit is measured in amperes (A). When 6.242×10^{18} electrons move past a point in 1 second (s), the current flow is said to be 1 ampere. The letter I is used to represent current. Scientific notation is often used to indicate the amount of current. Therefore, small amounts of current may be measured in milliamperes (mA) (0.001 ampere), and large amounts of current may be measured in kiloamperes (kA) (1000 amperes).

Voltage

Voltage is the difference in potential (charge) between two points. Expressed in another way, voltage is the amount of driving force or pressure applied to a circuit. The voltage in an electrical circuit is comparable to the pressure in a hydraulic circuit. The pressure in a hydraulic circuit is generally measured in pounds per square inch, while the voltage in an electrical circuit is measured in volts (V). Small voltages may be measured in millivolts (mV) (0.001 volt), while larger quantities are measured in kilovolts (kV) (1000 volts). The letter E is used to represent voltage.

Resistance

The resistance of a circuit is the circuit's opposition to the movement of electrons. The fewer free electrons that exist in the atomic structure of a material, the greater the opposition to the flow of current. Resistance may be compared to an orifice or small pipe that restricts flow in a hydraulic circuit. A resistor restricts or limits the amount of current flowing in an electrical circuit. The unit of measurement of resistance is the ohm. The symbol used for resistance is the Greek letter omega, Ω.

Ohm's Law

The relationship of current, voltage, and resistance is expressed in Ohm's law:

$$\text{Effect} = \frac{\text{cause}}{\text{opposition}}$$

$$\text{Current} = \frac{\text{voltage}}{\text{resistance}}$$

$$I = \frac{E}{R}$$

Thus, the current in a circuit is determined by dividing the applied voltage by the circuit resistance. Note that current varies directly with voltage and indirectly with resistance. The physical arrangement of resistors in a circuit affects the amount of opposition offered to the flow of current. Resistors may be connected in series, parallel, or a combination of series and parallel.

Series Resistors

When resistors are connected in series they have *one* point in common. The total resistance is equal to the sum of the individual resistors. Algebraically, this can be written

$$R_{tot} = R_1 + R_2 + R_3 + \cdots + R_n$$

The current in a series circuit is the same in each component of the circuit because the current must flow through each resistor in the series to get to the next resistor. This fact can be expressed as follows

$$I_{tot} = I_1 = I_2 = I_3 = \cdots = I_n$$

The applied voltage divides across each component in a series circuit in proportion to the resistance of the component. The greater the amount of resistance, the greater the voltage drop across that resistor. It is customary to use the letter E to symbolize a voltage applied to a circuit and the letter V to represent the voltage drop in a circuit. In a series circuit,

$$E = V_1 + V_2 + V_3 + \cdots + V_n$$

The *voltage divider rule* is used to calculate the voltage across each resistor as follows:

$$V_x = \frac{R_x V}{R_{tot}}$$

EXAMPLE 1.1

$$I = \frac{E}{R} = \frac{100}{R_1 + R_2} = \frac{100}{25} = 4 \text{ amperes}$$

$$V_1 = \frac{R_1 V}{R_{tot}} = \frac{5\,(100)}{25} = 20 \text{ volts}$$

$$V_2 = \frac{R_2 V}{R_{tot}} = \frac{20\,(100)}{25} = 80 \text{ volts}$$

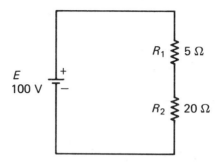

To check: The voltage applied to a circuit is dropped in the circuit, that is, $E = V$:

$$E = V_1 + V_2 = V$$

$$100 = 20 + 80$$

$$100 = 100$$

Parallel Resistors

When resistors are connected in parallel, they have two points in common. The total resistance of parallel resistors is equal to the reciprocal of the sum of the reciprocals of the individual resistors. R_{tot} of a parallel circuit is called the *equivalent resistance*, R_{eq}. Stated algebraically,

$$R_{eq} = \frac{1}{\dfrac{1}{R_1} + \dfrac{1}{R_2} + \dfrac{1}{R_3} + \cdots + \dfrac{1}{R_n}}$$

The current in a branch of a parallel circuit is equal to the equivalent resistance of the circuit divided by the resistance of that branch multiplied by the total current of the circuit; that is,

$$I_x = \frac{R_{eq}}{R_x} I_{tot}$$

EXAMPLE 1.2

100 volts is applied to each resistor.

$$I_1 = \frac{E}{R_1} = \frac{100}{5} = 20 \text{ amperes}$$

$$I_2 = \frac{E}{R_2} = \frac{100}{20} = 5 \text{ amperes}$$

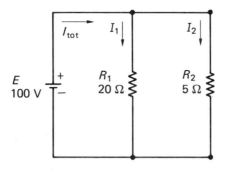

The current delivered to the circuit by the voltage source is the sum of the branch currents, $I_1 + I_2$, and equals 25 amperes.

The parallel configuration offers less resistance to current flow since each branch provides a path for current flow. The parallel circuit may be said to offer more *conductance*. Conductance is the reciprocal of resistance and is represented by the letter G. The unit of conductance is the mho which has the symbol ℧. For the parallel circuit of Example 1.2,

$$R_{eq} = \frac{1}{G_{tot}} = \frac{1}{\dfrac{1}{R_1} + \dfrac{1}{R_2}} = \frac{1}{\dfrac{1}{20} + \dfrac{1}{5}} = 4 \text{ ohms}$$

Series–Parallel Circuits

A series–parallel circuit contains some elements that are connected in series and others that are connected in parallel. Example 1.3 illustrates a series–parallel circuit.

EXAMPLE 1.3

$R_{tot} = R_1 + R_2//R_3$

$R_1 = 10$ ohms

$$R_2//R_3 = \cfrac{1}{\cfrac{1}{R_2} + \cfrac{1}{R_3}} = \cfrac{1}{\cfrac{1}{20} + \cfrac{1}{20}} = 10 \text{ ohms}$$

$R_{tot} = 10 + 10 = 20$ ohms

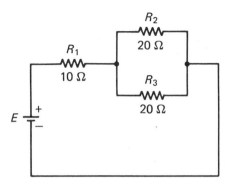

When resistors of equal value are connected in parallel, the equivalent resistance of $R_2//R_3$ may be found by dividing the number of parallel resistors into the value of the resistance. In this case, 20 divided by 2 equals 10 ohms. Obviously, this particular method is more convenient to use than the reciprocal method, but remember that it only applies when the parallel resistors are equal in value.

Color Coding and Standard Resistance Values

A wide variety of resistors are physically large enough to have their resistance value printed on them. However, carbon composition resistors are too small for this method of identification so a color coding system is used. Four color bands are printed on one end of the resistor and are read from the band closest to the end of the resistor toward the center. Each color represents a numerical value as indicated in Table 1.1. The first and second bands represent the first and second digits, respectively. The third band represents the multiplier, or the number of zeros that follow the second digit. The fourth band indicates the manufacturer's tolerance. If there is no fourth band the tolerance is ±20% of the rated value.

Table 1.1 Color Coding

0	Black	7	Violet	
1	Brown	8	Gray	
2	Red	9	White	
3	Orange	0.1	Gold	
4	Yellow	0.01	Silver	
5	Green	±5%	Gold	
6	Blue	±10%	Silver	Tolerance

EXAMPLE 1.4

Suppose the color bands of a resistor are yellow, violet, red, and gold. The resistance value is determined as follows:

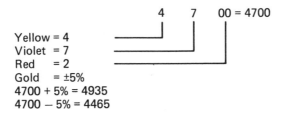

```
                              4    7    00 = 4700
Yellow = 4
Violet  = 7
Red     = 2
Gold    = ±5%
4700 + 5% = 4935
4700 − 5% = 4465
```

The actual resistance value should be between 4465 and 4935 ohms.

Occasionally a fifth band is used to indicate the failure rate of the resistor:

Yellow   0.001% per 1000 hours
Orange   0.01% per 1000 hours
Red   0.1% per 1000 hours
Brown   1.0% per 1000 hours

Kirchhoff's Voltage Law

Kirchhoff's voltage law states that "the algebraic sum of the potential rises and drops around a closed loop is zero." A closed loop is any continuous circuit through which current can flow from a point in one direction and return to that point from another direction. Stated algebraically, $\Sigma_Q V = 0$.

EXAMPLE 1.5

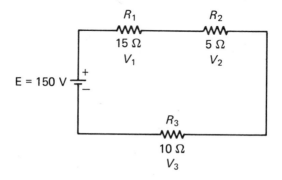

(a) Find R_{tot}.
Solution: $R_{tot} = R_1 + R_2 + R_3 = 15 + 5 + 10 = 30$ ohms

(b) Find I.

Solution: $I = \dfrac{E}{R_{tot}} = \dfrac{150}{30} = 5$ amperes

(c) Find V_1, V_2, and V_3.
Solution:

$$V_1 = IR_1 = (5)(15) = 75 \text{ volts}$$

$$V_2 = IR_2 = (5)(5) = 25 \text{ volts}$$

$$V_3 = IR_3 = (5)(10) = 50 \text{ volts}$$

(d) Verify Kirchhoff's voltage law.

$$\Sigma_Q V = E - V_1 - V_2 - V_3 = 0$$

$$E = V_1 + V_2 + V_3$$

$$150 = 75 + 25 + 50$$

$$150 = 150 \quad \text{(checks)}$$

Kirchhoff's Current Law

Kirchhoff's current law states that "the algebraic sum of the currents entering and leaving a node is zero." (A node is a junction of two or more branches.) Stated another way, the sum of the currents leaving a junction must equal the sum of the currents entering a junction.

EXAMPLE 1.6

Find I_3.

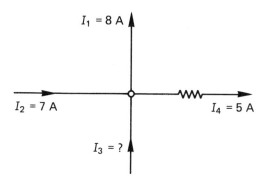

$I_1 = 8$ A

$I_2 = 7$ A $I_4 = 5$ A

$I_3 = ?$

Solution:

$$I_2 + I_3 - I_1 - I_4 = 0$$

$$I_2 + I_3 = I_1 + I_4$$

$$I_3 = I_1 + I_4 - I_2$$

$$I_3 = 8 + 5 - 7$$

$$I_3 = 6 \text{ amperes}$$

Conductors and Insulators

In addition to resistors, there are other components that affect the current in a circuit, for example, conductors and insulators.

Just as resistors oppose the flow of current in a circuit, conductors provide a nearly resistance-free path for current flow. Insulators restrict the flow of current to the intended path by isolating the circuit from adjacent material.

The ability of metals to conduct electricity, as well as their physical strength, makes them good conductors. However, not all metals conduct the same. For example, gold and silver are among the best conductors, but their high cost prohibit their extensive use. Copper is the most commonly used conductor because of its good conductivity and relatively low cost. Aluminum, which is less expensive than copper, is another metal commonly used as a conductor. However, aluminum does not conduct as well as copper (it is only about 60% as good).

As the ability of a material to conduct electricity (i.e., conductivity) decreases, its resistance increases. If the material does not conduct current, it is called an insulator. In actual practice, no insulator is perfect. There are different degrees of insulating properties, and insulators are known to fail at certain voltage levels. The term that is used to rate an insulator is called the *dielectric strength*. The dielectric strength is the breakdown point of an insulator. Some common materials used as insulators are mica, rubber, paper, pyrex, glass, and air.

A *semiconductor* belongs to the class of materials between the insulators and conductors. Germanium, selenium, silicon, and other similar compounds are semiconductors. These materials are vital ingredients for the production of transistors and other solid-state devices.

Capacitors and Inductors

When two conductors are placed side by side, separated by a nonconductive material, and connected across a battery, free electrons drift in the direction of the driving voltage. The battery in Figure 1.1 acts as a pump, removing electrons from one conductor and forcing them to the other.

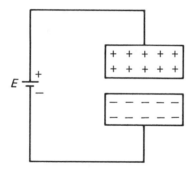

Figure 1.1

The pumping action or buildup of charge will continue until the *voltage* across the two conductors is equal to the charge of the battery. The size of the conductors will determine the amount of the charge that a capacitor can store.

If two different-sized capacitors, one large and one small, are connected to the same voltage source, voltage will build in each capacitor until it is equal to the battery voltage. However, the *capacity* of each is proportional to the size of the conductor plates. The larger conductor can store more of a charge than the smaller one because there are more free electrons in the larger conductor than in the smaller one.

The storage capacity of a capacitor may be compared to two sections of pipe, one 6 inches in diameter and the other 12 inches in diameter. If both sections were sealed and pressurized to 100 pounds per square inch, when opened, they would force the liquid out of the opening. The larger-diameter pipe would be able to supply more liquid than the smaller diameter pipe. Capacitors will work in the same manner. The larger the capacity, the greater the amount of charge that can be stored.

Inductors use the ability of electrical current to create a magnetic field. If a voltage is applied to a coil of wire, the current flowing in the coil will cause a magnetic field to develop. The more times the wire is coiled and the more current there is in the coil, the greater the strength of the magnetic field.

The property of a coil that opposes a change in the current flow is called *inductance*. The inductance of a coil depends on four factors:

1. The number of turns (windings) in the coil. Inductance is proportional to the square of the number of turns in the coil.
2. The diameter of the coil. The larger the diameter of the coil, the higher the inductance.
3. The permeability (ability to become magnetized) of the core material.
4. The length of the coil. The shorter the coil, the higher the inductance.

Power

Power is a rate of doing work, or work per unit of time. The unit for measuring power is the watt. Power in watts is equal to the product of the applied voltage and the current flow. Stated algebraically, $P = IE$.

Power in watts can also be expressed as follows:

$$P = \frac{E^2}{R} \quad \text{and} \quad P = I^2R$$

Alternating Current

The principles of alternating current (ac) are similar to those of direct current in that the same basic laws apply. The study of alternating current, however, is much more complex because of the continually changing amount and direction of the current and voltage. The components of an ac circuit cause a time period to be introduced between current and voltage; that is, current and voltage are out of phase. The material in this section does not present an in-depth study; rather, it is an overview to acquaint readers with the general characteristics of sinusoidal alternating current.

If a coil of wire is rotated through a magnetic field, a voltage is induced in the wire. The magnitude of the voltage depends on the amount of magnetic lines of force, called *flux lines*, that the conductor passes through in a given period of time. When 100×10^6 (one hundred million) lines of force are cut in 1 second, 1 volt is induced. The direction, or *polarity*, of the voltage is determined by the direction of motion of the wire with respect to the magnetic field. The illustration of Figure 1.2 shows the magnitude and direction of the induced voltage as a coil is rotated at a uniform speed through a magnetic field.

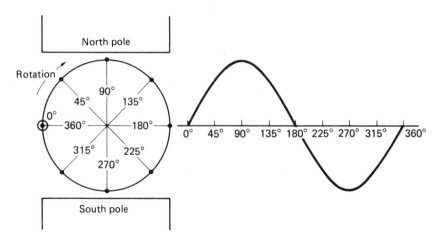

Figure 1.2

As the conductor passes through the 0° position, no voltage is generated because the conductor is moving parallel to the lines of force. Since no lines are cut, no voltage is generated. As the conductor rotates toward the 90° position the increase in the angle of the conductor with respect to the magnetic field causes more lines of force to be cut in each unit of time. That is, the value of voltage increases and is a function of the sine of the angle. At the 90° position, the conductor passes through the field cutting the maximum number of lines of force and a maximum voltage is generated. As the conductor moves toward the 180°

position, the value of generated voltage decreases. When the conductor passes 180°, the general direction of motion changes, and reverses the polarity of the voltage. The voltage values follow the first half-cycle but are of opposite polarity.

Definitions and Symbols

Amplitude or Peak Value
   The maximum value reached by a waveform.

Capacitive Reactance (X_C)
   Capacitive reactance, measured in ohms, is the opposition to a change in current flow provided by a capacitor. Capacitive reactance causes current to lead voltage by 90°: $X_C = 1/2\pi f_C$.

Cycle
   The portion of a waveform contained in one *period* of time.

Effective Value
   The value of voltage that occurs at 45° (0.707 times the maximum value).

Frequency
   The number of cycles that occur in 1 second.

Impedance (Z)
   The opposition to current flow in an ac circuit. It is a combination of resistance, inductance, and capacitance.

Inductive Reactance (X_L)
   Inductive reactance, measured in ohms, is the opposition to a change in current flow produced by a coil of wire. Inductive reactance causes current to lag voltage by 90°: $X_L = 2\pi f_L$.

Instantaneous Value
> The magnitude of a waveform at any instant in time.

Period (T)
> The time interval between successive repetitions of a periodic waveform.

Periodic Waveform
> A waveform that continually repeats itself after the same time interval.

Resistance
> The opposition of a circuit to the movement of electrons. Resistance in an ac circuit acts the same as resistance in a dc circuit.

Waveform
> The path followed by a voltage or current, such as the voltage waveform of Figure 1.1, plotted as a function of position. The position is also a function of the time that elapses as the conductor rotates through various positions.

Impedance in a *RL* Circuit

Figure 1.3 shows a circuit containing both resistance and inductance. The value of impedance for the circuit of Figure 1.3 can be solved graphically and is measured to be 5 units (ohms) long. A protractor can be used to measure the angle between resistance and impedance and is 53°. A mathematical solution for impedance involves right triangle trigonometry. Impedance can be found by the following formula:

$$Z = \sqrt{R^2 + X_L^2} \; \underline{/\tan^{-1}} \, (X/R)$$

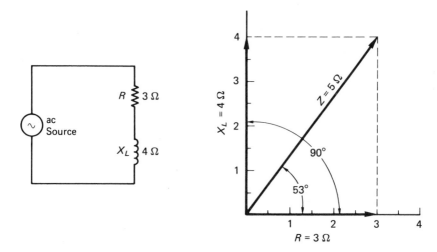

Figure 1.3

For Figure 1.3 the mathematical solution is

$$Z = \sqrt{3^2 + 4^2} \; \underline{/\tan^{-1} (4/3)}$$

$$Z = \sqrt{25} \; \underline{/\tan^{-1} (1.33)}$$

$$Z = 5 \; \underline{/53°}$$

The current in a RL circuit is out of phase with the source voltage by the impedance angle. The waveform of Figure 1.4 shows the phase relationship of current and voltage for the circuit of Figure 1.3.

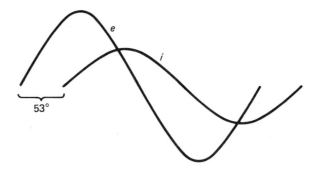

Figure 1.4

Impedance in a RC Circuit

Figure 1.5 shows a circuit containing both resistance and capacitance. The solution for impedance of a circuit containing resistance and capacitance is similar to the solution previously shown except that capacitance acts in the opposite direction.

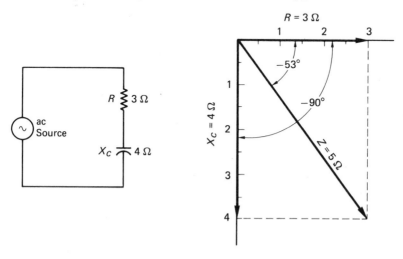

Figure 1.5

Inductive reactance is positive; capacitive reactance is negative. The following equation may be used to solve for impedance in a RC circuit:

$$Z = \sqrt{R^2 + X_C^2} \angle\tan^{-1}(X/R)$$

$$Z = \sqrt{3^2 + (-4)^2} \angle\tan^{-1}(-4/3)$$

$$Z = \sqrt{25} \angle\tan^{-1}(-1.33)$$

$$Z = 5 \angle{-53°}$$

The current in a RC circuit is out of phase with the source voltage by the impedance angle. The waveform of Figure 1.6 shows the phase relationship of current and voltage for the circuit of Figure 1.5.

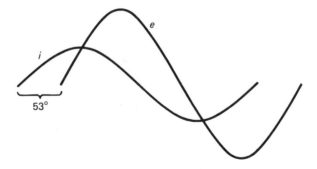

Figure 1.6

Impedance in a *RLC* Circuit

Figure 1.7 shows a circuit containing resistance, inductance, and capacitance. The value of impedance for the circuit of Figure 1.7 can be solved graphically and is measured to be approximately 8 units (ohms) long at an angle of 60°.

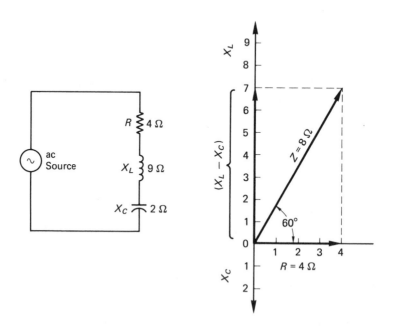

Figure 1.7

The mathematical solution is as follows:

$$Z = \sqrt{R^2 + (X_L - X_C)}\ \underline{/\tan^{-1}}\ [(X_L - X_C)/R]$$

$$Z = \sqrt{4^2 + (9 - 2)^2}\ \underline{/\tan^{-1}}\ [(9 - 2)/4]$$

$$Z = \sqrt{65}\ \underline{/\tan^{-1}}\ (7/4)$$

$$Z = 8.06\ \underline{/60.25^\circ}$$

It should be noted that the inductance is larger than the capacitance; therefore, the circuit is inductive.

To analyze ac circuits it is necessary to use complex numbers to represent current, voltage, and impedance; that is, numbers that give both magnitude and direction. The impedance values previously given are examples of complex numbers.

The value of an impedance of $5\underline{/53^\circ}$ may also be given as 3 ohms of resistance and 4 ohms of inductance. From the graph of Figure 1.8 it should be obvious that the point on the graph may be defined as 5 units away from the origin at a 53° angle, or as coordinates of 3 units along the x axis and 4 units along the y axis.

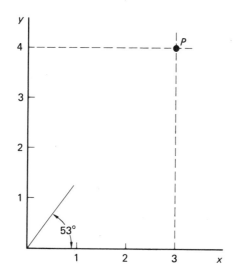

Figure 1.8

ance (R) in Ohm's law and the current of an ac circuit may be found. See Figure 1.9.

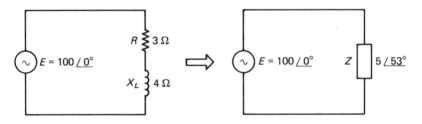

Figure 1.9

To solve for I, we use the following mathematical relationship:

$$I = \frac{E}{Z} = \frac{100\,\underline{/0°}}{5\,\underline{/53°}} = 20\,\underline{/-53°}\ \text{amperes}$$

The phase relationship of voltage and current is that shown in Figure 1.4.

Chapter 2

Direct Current (dc) Instruments

THE CONCEPTS DISCUSSED in Chapter 1 begin to take on more meaning as we move into practical applications. The instruments that are discussed in this chapter are used for measuring current, voltage, and resistance.

d'Arsonval Meter Movement

The d'Arsonval meter movement is probably the most commonly used meter movement. The electrodynamometer, which is used in the construction of a wattmeter, is discussed later in the chapter.

The d'Arsonval movement, shown in Figure 2.1, consists of a coil of wire wound on an iron core, mounted on bearings, and placed in the field of a permanent magnet. Springs are connected to the coil to prevent extreme rotation and to supply a current path to the coil. As current flows in the coil, a magnetic field is produced which interacts with the magnetic field of the permanent magnet, producing a torque. An indicator, or pointer, is fixed to the coil to show the amount of rotation. Since the strength of the permanent magnet is constant, the degree of rotation is a function of the current in the coil: The greater the current, the further upscale the indicator will move.

The meter movement is rated by current and resistance. A typical d'Arsonval movement rating is 1 mA (0.001 ampere) and 50 ohms. That is, the resistance of the coil is 50 ohms and a current of 1 mA in the coil will cause full-scale deflection of the indicator. Most meters are designed to indicate a zero reading on the left side of the meter face and to move upscale to the right as measured current increases. The face of a zero left meter is shown in Figure 2.2.

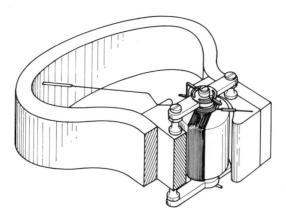

Meter movement parts

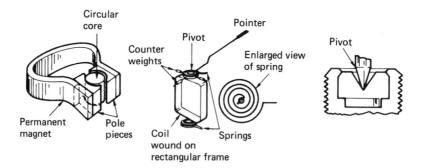

Figure 2.1

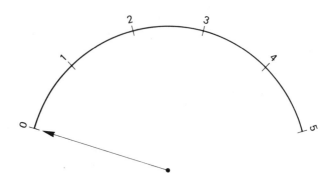

Figure 2.2

A meter movement may also be designed to indicate a zero reading in the center position and to move to the left or to the right depending on the current direction. The face of this type of movement is shown in Figure 2.3.

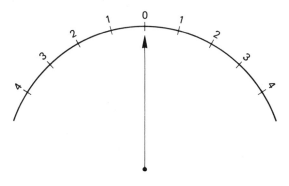

Figure 2.3

While only a very small current is permitted to flow in the meter coil, much larger currents can be measured by connecting a resistor in parallel with the coil. Suppose a current of 10 amperes is to be measured. By connecting a resistor in parallel with the coil, current will divide inversely in proportion to the resistance. If the shunt resistance is very carefully selected, 1 mA of current will flow in the meter coil when 10 amperes flow in the circuit.

The resistance value of the shunt resistor (commonly called a *meter shunt*) is determined to permit 1 mA of current to flow in the meter coil when the maximum current to be measured flows in the circuit. If 0.001 ampere flows in the meter coil and 9.999 amperes flow in the shunt resistor, then the current is the sum of the two currents, or 10 amperes. This relationship can also be expressed as

$$I_{shunt} = I_{total} - I_{coil}$$

$$10 - 0.001 = 9.999 \text{ amperes}$$

Also, since the meter coil and the shunt resistor are connected in parallel, their voltages must be equal:

$V_{coil} = V_{shunt}$

$V_{coil} = I_{coil} R_{coil} = 0.001\ (50) = 0.05\ \text{volts}$ or   50 mV

The resistance of the shunt may be found by Ohm's law, $R = E/I$:

$$R_{shunt} = \frac{V_{shunt}}{I_{shunt}} = \frac{0.5}{9.999} \simeq 0.005\ \text{ohms}$$

EXAMPLE 2.1

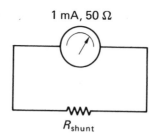

1 mA, 50 Ω

R_{shunt}

Find the value of shunt resistance to be used in the above meter circuit if the maximum current to be measured is 25 amperes.

$V_{shunt} = V_{coil} = 0.001\ (50) = 0.05\ \text{volts}$

$I_{shunt} = I_{total} - I_{coil} = 25 - 0.001 = 24.999\ \text{amperes}$

$$R_{shunt} = \frac{V_{shunt}}{I_{shunt}} = \frac{0.05}{24.999} \simeq 0.022\ \text{ohm}$$

If several values of maximum current are to be measured by the same meter, the circuit of Figure 2.4 can be used. Resistance values for each of the meter shunts are determined as illustrated above.

Care must be taken to be certain that an ammeter is connected in series with the current to be measured. If a multirange meter is used, the selector switch must be set on a range that will accommodate the current. For example, if the selector switch is set on the 1 ampere range and 10 amperes are applied to the meter, the meter will be damaged. Furthermore, correct polarity must be observed for upscale deflection. The positive

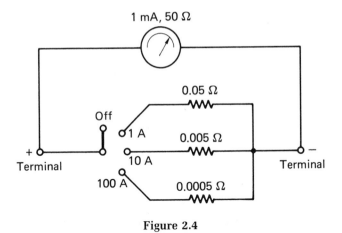

Figure 2.4

and negative terminals on the meter must correspond to the polarities of the current to be tested.

Voltmeter

If the d'Arsonval meter movement is used for a voltmeter and a resistor is connected in series with the coil, the applied voltage will divide across the coil and the added series resistor. Figure 2.5 illustrates the resistance connections for a voltmeter.

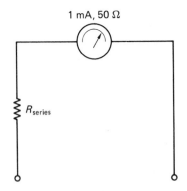

Figure 2.5

If the maximum voltage to be measured is 10 volts, the value of R_{series} must be selected so that when 10 volts are applied to the meter, 0.05 volt will be dropped on the meter coil. Applying Kirchhoff's voltage law around the closed loop of Figure 2.5 yields the following equation:

$$+\ 10 - (0.001)\ (R_{series}) - 0.001\ (50) = 0$$

$$R_{series} = \frac{10 - (0.001)\ (50)}{0.001}$$

$$R_{series} = 9950\ \text{ohms}$$

A multirange voltmeter may be constructed if various resistance values are connected for selection as desired. The calculations for the value of series resistance to be used for each range are the same as those illustrated. Figure 2.6 shows the connection pattern for a multirange voltmeter.

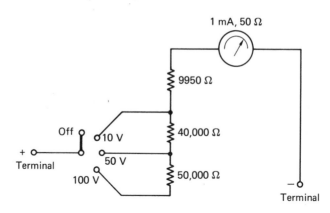

Figure 2.6

When a value of 10 volts is selected, R_{series} = 9950 ohms; when 50 volts is selected, R_{series} = 40,000 + 9950 = 49.9K ohms; when 100V is selected, R_{series} = 50,000 + 40,000 + 9950 = 99.5K ohms. When measuring voltage, the voltmeter must be connected in parallel with the voltage to be measured. For an upscale deflection, the polarity of the voltage to be measured must correspond to the polarity indicated on the meter.

The described voltmeter is for use with direct current sources. If connected to an alternating current source, the meter would not give a meaningful reading and would be damaged if left connected for an extended period of time. However, the meter can be modified to read ac voltages by including a rectifier circuit in the meter.

Series Ohmmeter

The series ohmmeter also uses the d'Arsonval meter movement. A fixed-value resistor, an adjustable resistor, and a battery are connected in series. The resistance to be measured is connected across the circuit as shown in Figure 2.7.

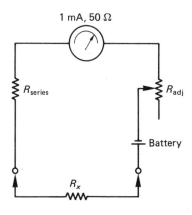

1 mA, 50 Ω

R_{series}

R_{adj}

Battery

R_x

Figure 2.7

The battery in the meter causes a current to flow in the circuit, including the meter coil. The current is a function of the coil resistance, R_{series}, R_{adj}, and R_x, and the battery voltage. Before connecting the meter to R_x, the meter leads should be connected together and R_{adj} adjusted to obtain a full-scale reading. This is obtained when R_x equals zero. Also, an infinite reading appears when the meter leads are not connected. Any resistance that is to be measured must then be between infinity and zero, moving upscale from infinity toward zero.

The value of the series resistor may be determined as follows:

$$R_{series} = \frac{E}{CS} - R_m - \frac{R_{adj}}{2}$$

where   E = the voltage reading of the battery
        CS = the current sensitivity of the meter (the current required to operate the meter to full scale)
        R_m = the resistance of the meter coil
        R_{adj} = the adjustable resistor

The ohmmeter scale is nonlinear because increasing values of resistance to be measured cause an increasingly lesser amount of current to flow in the meter coil. This is a result of the change in the ratio of the unknown resistor to the other resistors.

An ohmmeter must never be connected to an energized circuit. The circuit voltage applied to the meter in this manner may be sufficient to damage the meter. At any rate, a meaningful reading cannot be obtained.

Electrodynamometer

The electrodynamometer type of meter movement is more complex than the d'Arsonval movement. It can be used in each of the meters previously discussed and, in addition, can be used in the construction of a wattmeter. In the electrodynamometer movement, the magnetic field is produced by a separate winding instead of using a permanent magnet as in the d'Arsonval movement. The magnetic fields of the stationary winding and the moving coil interact to determine the movement of the indicator which is connected to the moving coil. Since a reversal of current reverses the polarity of both magnets, the indicator always moves upscale, regardless of the applied polarity. For this reason, the electrodynamometer movement can be used to measure either ac or dc values. With the electrodynamometer movement it is also necessary to use resistors to limit the current in the coils.

When used as a wattmeter, the current in the stationary coil is a function of line current and the current in the moving coil is a function of the line voltage. The voltage coil must be connected in parallel, and the current coil in series with the circuit or component being measured. The circuit for a wattmeter is shown in Figure 2.8.

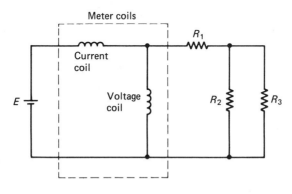

Figure 2.8

Megger

The megohmmeter, commonly called a megger for short, is used to measure very high resistance values. It is primarily used to test the insulation of conductors. To measure high resistance values, a high voltage is applied, either by the use of a hand-cranked generator or an electronic power supply.

In use, one lead of the megger is attached to the device being tested and the other is attached to a ground. When the voltage is applied, any breakdown in the insulation of the circuit being tested is indicated on the scale. Some common megger test voltages are 250, 500, and 1000 volts. A device being tested should be checked on the next highest voltage above its rating.

Care should be taken to follow instructions very closely when checking circuits containing solid-state devices. These devices are very sensitive to the high voltage and can be destroyed by the voltage generated by the megohmmeter.

Clamp-On Ammeters

Clamp-on ammeters are used to check the current in a circuit, without being physically connected in the circuit. They are available in many different sizes and can be used for either ac or dc current. Clamp-on ammeters are convenient to use in the field since the circuit does not have to be opened to take a current reading.

Infrared or Thermal Scanners

Infrared or thermal scanners are generally hand-held devices that produce an image of a component showing temperature variations. This is effective in spotting worn or loose connections and components in industrial circuits. These devices can be used by inspectors who aim the device at a control panel. The parts of the panel operating at a higher temperature than desired will be highlighted on the image.

When the inspector spots a high-temperature connection, he or she can record the defect for correction on the next down period for the equipment.

These devices can thus prevent the shutdown of electrical equipment for standard inspections.

Phase Sequence Indicator

Phase sequence indicators come in two styles—lights and meters. In the lighted variety, a sequence of lights goes on for the phase sequence being read, while the meter indicates which phase direction it is reading.

Phase sequence indicators are used when connecting and disconnecting equipment are commonplace. When connected improperly, a phase sequence device will operate in a reverse direction. This can range from annoying to catastrophic in certain operations. A phase sequence indicators thus ensures correct operation.

Rotation Tester

This device is used during the installation of a motor to determine the direction of rotation of the motor once it is installed. The shaft is mechanically rotated in the desired direction and the meter indicates if that is the direction in which the motor will rotate. To check the supply, the phase sequence indicator, mentioned previously, should be used.

Miscellaneous Testers

Testers come in a variety of types for many purposes. Some of the common types of small tester include hand tool and receptacle testers.

A hand tool tester has a receptacle into which the tool is plugged. Once the tool is plugged in, you are able to test the tool for shorts, open circuits, and grounds. This will prevent tool failure and injury to the operators.

A receptacle tester is used to test the proper connection pattern of the receptacle. If wires are connected incorrectly, the lighting arrangement on the receptacle indicates a fault.

Chapter 3 Devices, Symbols, and Circuits

A WIDE VARIETY OF DEVICES are used in both manual and automatic control. The purpose of this text is not to discuss each component at length, but rather to present the basic concepts of control circuits and components to enable the reader to apply the concepts to any device or circuit. As was discussed in Chapter 1, a circuit is an arrangement of components to form a closed loop.

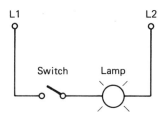

Figure 3.1

The circuit of Figure 3.1 consists of a toggle switch and a light bulb. As is evidenced by the sketch, called a *circuit diagram*, symbols have been used to represent the actual components. If the switch is closed, and this must be done manually, the circuit is completed from line 1, through the switch, through the lamp, to line 2. The completed circuit permits current to flow through the lamp. No matter how complex or involved a circuit may become, this basic principle must apply. That is, symbols are used to represent the actual device in the control circuit, and anyone who wants to read and understand the circuit must know the symbols and how the represented device functions in the circuit. Following are some of the common symbols. The devices will be defined further as used in circuits.

Symbol	Description
	Single-pole, single-throw switch.
	Three-pole, single-throw switch. Note that the blades are mechanically tied together. This is indicated by the dotted line.
	Lamp.
	Shunt coil, the operating coil of a relay or contactor.
	Series coil, often used in overload relays.
	Normally open push button (normal means at rest or unactivated) may be momentary engaged or maintained.
	Normally closed push button.
	Combination push button, one side normally open and one side normally closed.
	Limit switches, normally open and normally open–held closed.
	Limit switches, normally closed and normally closed–held open; used to convert mechanical motion into an electrical control signal.

Symbol	Description
	Normally open and normally closed foot-operated switches used where the process requires that the operator have both hands free.
	Pressure or vacuum switches.
	Liquid level switches (float switches).
	Temperature-actuated switch.
	Flow switches (air, water, oil, etc.).
	Instant operating contacts.
ON	Contacts of a timing relay. Contact action is retarded on energizing the operating coil. Designed to control a preset time period.
OFF	Contacts of a timing relay. Contact action is retarded on deenergizing the operating coil.
REV FOR	Master switch contacts operated by a rotating cam. The X at the intercept of the vertical dotted line indicating the controller position and the control wire demonstrates that the contact is closed at that position.

Symbol	Description
—(M)—	Single-phase motor.
T1 / T2 / T3	Three-phase squirrel cage motor. The most common and reliable industrial motor, it is used extensively in constant-speed applications.
A1 —(A)— A2	Armature of a dc or universal motor. It is used extensively in applications where speed control is necessary.
F1 ⌇⌇⌇ F2	The shunt field of a dc motor. The magnetic flux produced is independent of motor load.
S1 ⌇⌇ S2	Series field of a dc motor. The magnetic flux produced is dependent on motor load.
H1 H3 H2 H4 / X1 X2	Dual voltage transformer used in control circuits.

The circuit of Figure 3.1 could just as easily have been constructed to control a motor. Assume it is desired to start and stop a motor used to operate a ventilating fan in the roof of a large storage warehouse. A circuit, consisting of a switch large enough to handle the motor current and the motor, could be constructed as shown in Figure 3.2.

With the circuit of Figure 3.2, either the operator is required to go to the location of the power supply to start and stop the fan motor each time such actions are required, or wires are run to a convenient operating location. Obviously, it would be

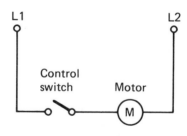

Figure 3.2

inconvenient for the operator to go to the power source. It would also present a safety hazard in the case of a very large motor. To run very large wires to a remote or desired control location could be expensive, and adding the extra length of wire to the system would introduce needless power loss. The solution is to install a *remote control system*; that is, to construct a circuit that permits an operator in one location to close a switch in another location. This action is achieved by the use of a contactor, or *relay*.

In the construction of a relay, a coil of wire is wound on a pole piece to produce a magnet, and the magnet is used to operate a switch. The switch is called contacts or tips. This principle permits an operator in one location to control a very small current and to operate a switch in another location controlling a very high current. See Figure 3.3.

It should be obvious that our total circuit now consists of a control circuit, in which the coil and the initiating device are located, and a power circuit, containing the motor and the load switch. The control circuit can be constructed of a very light wire because the current is only the amount of current necessary to produce a magnetizing force for operating the relay. Many control circuits operate on less than 1 ampere. The low current and small wire make control from a remote location more practical. The current in the power circuit can be and usually is many times the current in the control circuit. In many instances, the power circuit also operates on a higher voltage than the control circuit, as is indicated in the circuit of Figure 3.3 Note that the circuits are electrically separate.

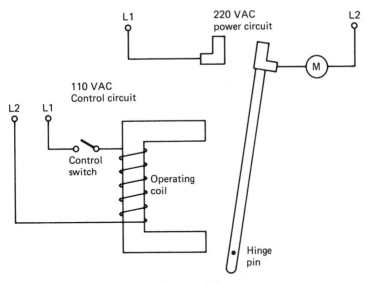

Figure 3.3

Relay circuits can also be used to establish sequence control. The circuit of Figure 3.4 illustrates a simple control circuit used to control three lamps connected to light in a sequence, red, green, and amber. The operation of each one depends on the operation of the previous ones.

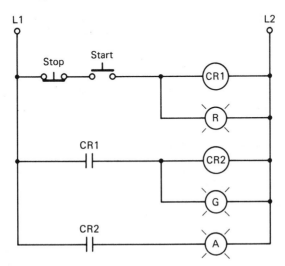

Figure 3.4

The electrical operation of the circuit shown in Figure 3.4 is as follows: Pressing the normally open START button will permit current to flow from L1 through the normally closed STOP button, through the START button, CR1 coil, and the red lamp, to L2. Note that the CR1 coil and the red lamp are in parallel and thus have the same voltage applied. The current in the CR1 coil will produce a magnetizing force and the relay will operate and close CR1 contacts. Another circuit is set-up, and current will flow from L1 through CR1 contacts, through the CR2 coil and the green lamp to L2. Now the CR2 coil becomes energized (magnetized) and operates its contacts. The final path (circuit) is completed from L1 through CR2 contacts and the amber lamp to L2. While a time delay did not occur between turning on the lamps, each lamp was turned on in sequence. The principles explained here can be expanded to control complex operations whose output must depend on some previous action, a fluid level, flow, pressure, position, or some physical condition that can be used to operate a control device.

Manual Control

Manual control involves the type of control that was illustrated in Figures 3.1 and 3.2. Both circuits require "hand" control at the location of the controller. Manual control lends itself to applications where control requirements are simply to start and stop an operation. Manual control can be designed to provide overload protection, speed control, reversal of direction, and very little else. Examples of manual control are found in work-shops, where grinders, drill presses, lathes, and other machine tools are used. Again, manual control can be characterized by the fact that an operator must initiate the control action.

Semiautomatic Control

Semiautomatic control involves the use of magnetic relays or contactors and an arrangement of one or more manually operated

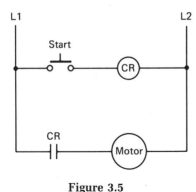

Figure 3.5

initiating devices such as a push button. The circuit of Figure 3.5 constitutes a semiautomatic control circuit. As in manual control, semiautomatic control requires that an operator initiate any required change in the operation. The principal advantage of semiautomatic control over manual control is the capability of installing a remote control station.

Automatic Control

Automatic control involves the use of magnetic relays or contactors and an arrangement of one or more initiating devices that operate automatically. For example, the contacts of a float switch may be arranged to close when the fluid in a tank rises to a certain level. The switch will automatically close each time the predetermined high fluid level is reached. A circuit can be constructed whereby the closing of the float switch will start a pump motor each time the high level is reached. It should be further noted that at a predetermined low level, the float switch will open. Such an arrangement could be connected to control a sump pump and is shown in Figure 3.6.

The contacts in a float switch can be very small, and usually are, because the current through them is only the amount necessary to magnetize the coil of the relay. The size of the motor is determined by the load and the contacts of the relay are selected according to the load current.

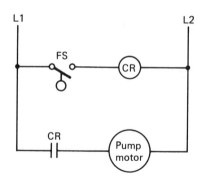

Figure 3.6

If the current in the load circuit is very large, an additional relay may be placed in the circuit to increase current handling capabilities. A magnetic switching device whose contacts are placed in a control circuit is called a relay, and a device whose contacts are placed in the load circuit is called a contactor. It may be said that a relay controls a contactor and a contactor controls a load. In the circuit shown in Figure 3.7, the float switch is the initiating device (called a primary pilot control component) and could be any type of control component depending on the operation. The closing of the float switch will cause a current to flow in the coil of the relay, CR1. When CR1 contacts close, a current will flow in the coil of the contactor, C.

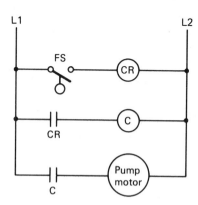

Figure 3.7

The contactor is larger than the relay and is designed to handle much more current. The closing of C contacts will then start the motor. It is important to note that a microswitch can be used to signal the control circuit to perform its function while a large current in the power circuit can achieve the desired function.

Wiring Diagrams

A wiring diagram shows as closely as possible the physical location of all parts of a circuit. All poles, terminals, contacts, and coils are shown for each device. A wiring diagram is helpful in the initial wiring of a circuit or when it is necessary to trace wires when troubleshooting. The actual sequence of operation is difficult to obtain from a wiring diagram.

Elementary Diagram

An elementary diagram gives an easy to understand sequence of operation of an electrical system without regard to actual component location. Control components are shown between a pair of vertical lines that represent the power supply lines. The elementary diagram is also called a ladder diagram, a line diagram, and a schematic.

Wire Numbers

To assist in troubleshooting elementary diagrams, identify each control wire. A numbering system is used whereby each wire or wire segment between components will have a separate number. In other words, each time a control wire is interrupted by a component a new number is used.

Location Numbers

Lines of control are numbered in the left margin of the drawing. The numbers in the right margin indicate the location on the drawing of the contacts operated by a particular coil. When the number is underlined, the contact is normally closed; otherwise it is normally open. A number is often placed at the contact to show the location on the drawing of its operating coil.

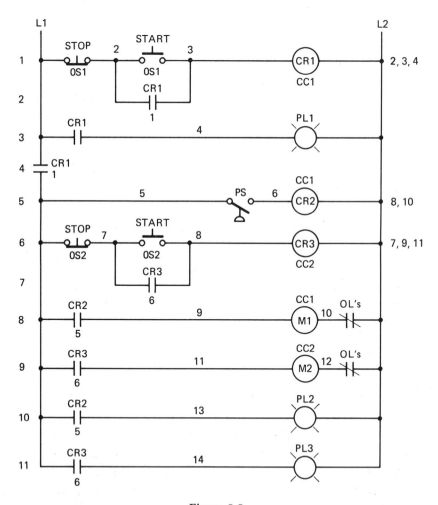

Figure 3.8

In complex systems where more than one control center is used, a number is placed under the coil to indicate in which control center the relay is located. Figure 3.8 illustrates the various numbers.

Three-Phase Motor Starters

Motor Starters

THE MOST COMMON AND RELIABLE INDUSTRIAL MOTOR is the three-phase squirrel cage motor. The purpose of this section is to explain the various starting circuits for the squirrel cage motor. The motor characteristics and the required three-phase power supply are explained later. The three-phase squirrel cage motor is usually started by connecting each of the three power supply lines directly to the motor terminals. This method of starting is called across-the-line starting. The contacts of the motor starter may be closed manually or magnetically as determined by the needs of the operation. This section will deal with magnetic starters.

The magnetic motor starter consists of primary pilot control devices as needed for the operation, an operating coil, three main contacts (one for each of the power lines), the required number of auxiliary contacts, and overload relays.

The circuit may be either two-wire or three-wire depending on the needs of the operation. The power circuit is usually of a higher voltage because of power advantages, while the control circuit voltage is lower for safety considerations. A control transformer is used to transform control circuit voltage to the desired voltage. A typical motor starter using the basic components is shown in Figure 4.1.

The operation of the circuit of Figure 4.1 is as follows: Pressing the START button will establish a current path from X1 of the control transformer, through the STOP button, the START button, the operating coil, the overload relay contacts, and the control circuit fuse, to X2 of the control transformer. Note that the high-voltage side of the transformer is connected to two of the power circuit wires, L1 and L2, applying 480 volts to the primary. Because of the turns ratio (4:1) of the primary winding to the secondary winding, the control voltage is reduced to 120 volts. While the control circuit receives its supply from the power circuit, it is connected to the power supply only through

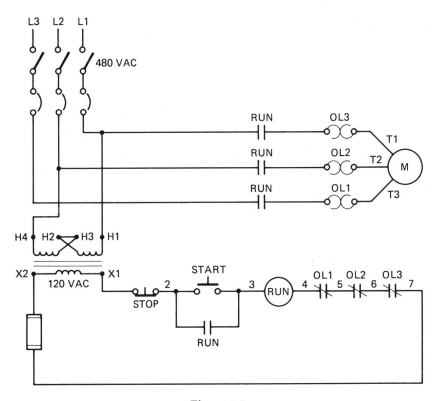

Figure 4.1

the mutual flux of the transformer. Two separate circuits exist—the power circuit and the control circuit.

Once the START button is pressed, the control circuit is completed and the operating coil, RUN, energizes and closes all its contacts. The RUN contacts in the power circuit connect the line leads to the motor leads and the motor starts. Simultaneously, the RUN contacts parallel to the START button close to provide a current path around the START button. The START button may now be released. The circuit will be maintained by the RUN auxiliary contacts in the control circuit. Pressing the STOP button will interrupt current to the RUN coil and all RUN contacts will open and disconnect the motor from the line. An overload condition, which causes the overload contacts to open, or a power failure would also cause the starter to drop out.

Reversing Control

To achieve reversal of the direction of shaft rotation for a squirrel cage motor, two motor starters are connected so that one starter is wired to connect L1, L2, and L3 to T1, T2, and T3, respectively, for forward operation, and the other starter connects L1 to T3, L2 to T2, and L3 to T1, for reverse operation. The starters are interlocked both mechanically and electrically to ensure that both starters cannot be operated simultaneously. The circuit of Figure 4.2 illustrates the power circuit for a reversing starter.

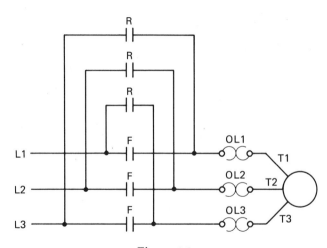

Figure 4.2

The control circuit for Figure 4.2 is designed to enable the operator to select either forward or reverse rotation of the motor shaft. Figure 4.3 illustrates a method of control.

The circuit of Figure 4.3 operates as follows: Pressing the FOR button will complete a circuit from X1 through the STOP button, the FOR button, R auxiliary contacts, the overload contacts, and the fuse to X2. The F coil will energize and operate all its contacts. The normally open F contacts parallel to the FOR button will close to provide a maintaining circuit and the FOR button may be released. The circuit will remain energized. The normally closed R auxiliary contacts in the R coil circuit will

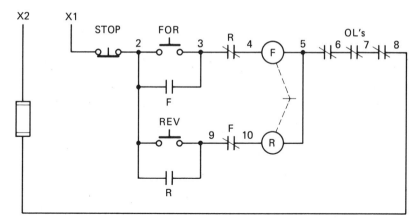

Figure 4.3

open to prevent reverse operation while operating forward. The dotted line between the coils indicates the mechanical interlock between the starters. The normally closed auxiliary contacts provide an electrical interlock. The circuit will continue to operate until the STOP button is pressed or a fault occurs to shut down the operation.

Another method of reversing control uses electrically interlocked push buttons and is shown in Figure 4.4. Note that the forward circuit must be made through the normally closed side of the reverse push button, and the reverse circuit must be made through the normally closed side of the forward push button. For either operation, pressing the opposite push button will shut down the operation.

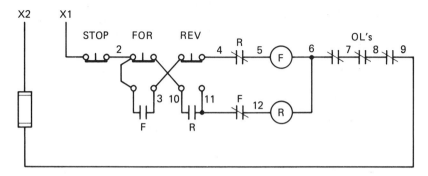

Figure 4.4

Definition of Terms

Duty Cycle Rating

Most motors have a continuous duty rating to permit continuous operation at a rated load. However, motors may be rated as intermittent duty, periodic duty, or varying duty, and must be turned off and allowed to cool after a fixed operating time.

Full-Load Current

The current required to produce full-load torque.

Jogging

The starting and stopping of a motor at frequent intervals.

Motor Controller

A device that controls some or all of the following functions: starting, stopping, overload protection, overcurrent protection, reversing, changing of speed, sequence control, and running/jogging,

Motor Speed

The shaft speed of the three-phase squirrel cage motor is determined by the frequency of the supply voltage and the number of poles in the motor. A two-pole motor runs at about 3550 rpm on 60 cycles per second. A four-pole motor, on the same frequency, will run at about 1750 rpm. The speed of a squirrel cage motor may be determined by the following formula:

$$\text{rpm} = \frac{\text{cycles per second}}{\text{pairs of poles}} \times 60 - \text{slip}$$

where slip is the difference between the speed of the rotating magnetic field and the speed of the rotor.

Overcurrent Protection

A fusible disconnect or circuit breaker used to protect the branch circuit conductors, control devices, and the motor from grounds and short circuits. The overcurrent protection device must be capable of carrying the starting current to exceed 400% of the motor full-load current.

Overloads

Any excessive amount of current drawn by a motor. Overloads on a motor may be mechanical or electrical.

Plugging

The instant reversal of a motor. Damage to the driven machinery can result if plugging is applied improperly.

Sequence Control

The control of separate motors to operate in a predetermined pattern.

Service Factor

The amount of overload that may be permitted without causing significant deterioration of the insulation on a motor. For example, if a 10 hp motor has a service factor of 1.15, the motor can safely be subjected to a 11.5 hp load.

Starter

The simplest form of a controller.

Starting Current or Locked Rotor Current

The current flow in the motor at the instant of starting. This current can be four to ten times the full-load current of the motor. The most common locked rotor current is about six times the full-load current. Such a motor will start with a 600% overload.

Torque

The twisting force produced by the motor. Torque is related to horsepower by the following formula:

$$\text{Torque} = \frac{\text{horsepower} \times 5252}{\text{revolutions per minute}}$$

Torque is in foot-pounds (ft-lb).

Overload Protection

The effect of an overload is an increase in the temperature of the motor conductors. If sustained high-overload conditions occur, the insulation of the motor can be damaged. Because of the time required for heat to transfer, the larger the overload the faster the temperature will increase to the damaging point. An inverse relationship exists between current and time. A large overload can cause instantaneous damage, where a small overload over a short duration will cause little or no damage. However, a small overload sustained for a long period of time can be just as damaging as a large overload. The relationship between overload current and time is shown in the heat transfer curve of Figure 4.5.

The curve indicates the time in minutes that an overload condition may exist without overheating the motor. For example, a 300% overload would reach its permissible temperature limit in 3 minutes.

Overload Relay

When an overload exists, the excessive current may be used to activate the overload protection device by placing a heating element in series with the motor and using the heat that is produced to activate a normally closed contact. The contact is connected to turn off the control circuit. Figure 4.6 shows the mechanism of a solder-pot-type thermal overload relay.

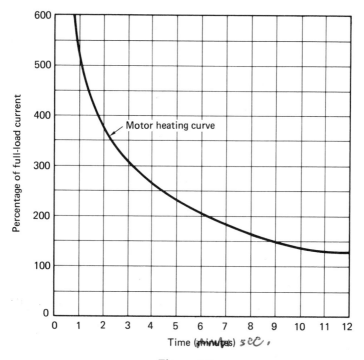

Figure 4.5

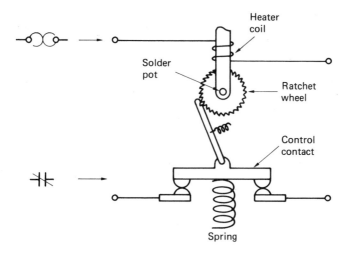

Figure 4.6

Overload relays may be thermal or magnetic. As the names imply, thermal relays operate on the heat that is produced by excessive current, while magnetic relays operate on the magnetic attraction produced by the current. Thermal overload relays are available in two types—melting alloy and bimetallic strip. In the melting alloy type, the heat produced by an overload current causes solder to melt, permitting a spring-loaded ratchet wheel to turn and open a set of contacts. In the bimetallic strip type, the heat produced by overload current causes a deflection of the bimetal and opens a set of contacts. The heating elements of each type must be matched to the load. Each type is equipped with optional manual or automatic reset mechanisms.

Figure 4.7 shows the heat transfer curve of a motor and typical thermal overload trip characteristics. The curves show that no matter how much current the motor draws, the overload relay will provide protection without needless tripping. Observe that the overload relay will always trip at a safe value.

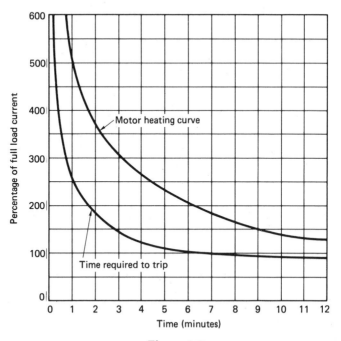

Figure 4.7

Magnetic Overload Relay

The magnetic overload relay has a movable core placed inside the relay coil. As current in the coil increases, the magnetic attraction also increases. When the core has moved sufficiently, a set of contacts will open. Time delay is provided by adjusting the position of the core. Movement of the core is retarded by a dash pot fluid that must pass through an orifice as the core moves. An inverse time characteristic is produced; that is, the greater the current, the less the amount of time required to open the contacts. Inverse time element overload relays are labeled ITE on control drawings. Magnetic overload relays may also be designed to trip instantaneously by eliminating the dash pot fluid to permit instantaneous motion of the core upon an overload condition. Instantaneous overload relays are labeled INST on control drawings.

Selection and maintenance of overload relays are extremely important to the life of a motor. The manufacturer's current rating for a particular operation should be adhered to strictly. See Figure 4.8. Repeated tripping of overload relays may be caused by mechanical misalignment of driven machinery, bad bearings, or simply an oversized load. The solution to repeated tripping of overload relays is not to replace the heating element but to remedy the overload condition. Probably the greatest single cause of motor failure is the frequent practice of installing a larger-sized heating element than the one required. Refer to Figure 4.9 for approximations of full load currents for three-phase squirrel cage induction motors.

When one line of a three-phase circuit opens, the motor loses its three-phase power supply and does not operate. This condition is called *single phasing*. The current flow in the other two lines becomes excessive and the overload relays trip to prevent damage to the insulation in the motor. Two overhead coils placed in lines 1 and 3 provide *standard* overload protection. Three overload coils, one in each of the three lines, provide *full* overload protection.

NEMA SIZE	Volts	Maximum Horsepower Rating— Nonplugging and Nonjogging	Maximum Horsepower Rating— Plugging and Jogging	Continuous Current Rating— Amperes at 600 volts Maximum	Service— Limit Current Rating*
00	110	0.75	—	9	11
	208–220	1.5	—	9	11
	440–550	2	—	9	11
0	110	2	1	18	21
	208–220	3	1.5	18	21
	440–550	5	2	18	21
1	110	3	2	27	32
	208–220	7	3	27	32
	440–550	10	5	27	32
2	110	7.5	—	45	52
	208–220	15	10	45	52
	440–550	25	15	45	52
3	110	15	—	90	104
	208–220	30	20	90	104
	440–550	50	30	90	104
4	110	—	—	135	156
	208–220	50	30	135	156
	440–550	100	60	135	156
5	110	—	—	270	311
	208–220	100	75	270	311
	440–550	200	140	270	311
6	110	—	—	540	621
	208–220	200	150	540	621
	440–550	400	300	540	621
7	110	—	—	810	932
	208–220	300	—	810	932
	440–550	600	—	810	932
8	110	—	—	1215	1400
	208–220	450	—	1215	1400
	440–550	900	—	1215	1400

Figure 4.8. Electrical ratings for ac magnetic starters. (*Per NEMA Standards, paragraph 1C 1–21A.20, the service-limit current represents the maximum rms current, in amperes, which the controller may be expected to carry for protracted periods in normal service. At service-limit current ratings, temperature rises may exceed those obtained by testing the controller at its continuous current rating. The ultimate trip current of overcurrent (overload) relays or other motor protective devices should not exceed the service-limit current ratings of the controller.)

| | Full-Load Current (Amperes)* at | | | | |
Horsepower	110 volts	220 volts	440 volts	550 volts	2300 volts
0.5	5.0	2.5	1.3	1.0	—
0.75	5.4	2.8	1.4	1.1	—
1	6.6	3.3	1.7	1.3	—
1.5	9.4	4.7	2.4	2.0	—
2	12.0	6.0	3.0	2.4	—
3	—	9	4.5	4.0	—
5	—	15	7.5	6.0	—
7.5	—	22	11	9.0	—
10	—	27	14	11	—
15	—	38	19	15	—
20	—	52	26	21	5.7
25	—	64	32	26	7
30	—	77	39	31	8
40	—	101	51	40	10
50	—	125	63	50	13
60	—	149	75	60	15
75	—	180	90	72	19
100	—	246	123	98	25
125	—	310	155	124	32
150	—	360	180	144	36
175	—	—	—	—	—
200	—	480	240	195	49

Figure 4.9. Full-load currents for three-phase squirrel cage induction motors. (*Approximations of full-load currents derived from average values for representative motors of their class.)

Reduced-Voltage Starters

Reducing the voltage applied to the motor terminals during the starting period is sometimes necessary because of limited capabilities of the power distribution system and to minimize the shock to the driven load. Typical applications are those where belts slip, chains are broken, or drive cases are damaged because of a sudden start and also where power disturbances are objectionable to other operations. Figure 4.10 illustrates the relationship of the full-load motor current and the motor speed from starting to full speed. Two things stand out. The starting

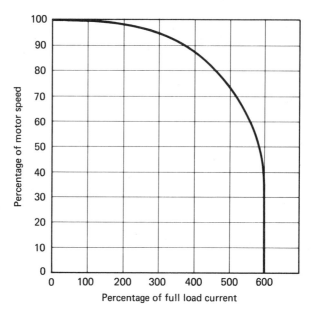

Figure 4.10

current is high and remains high throughout most of the starting period. Note that most of the reduction in current occurs during the final portion of the curve.

Typical Starting Methods

The following one-line diagrams illustrate various methods of achieving reduced voltage starting:

Primary Resistance Starting

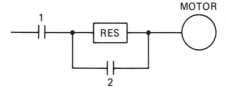

In primary resistance starting, a resistor is connected in the motor circuit during the starting period and removed for run.

Reactor Starting

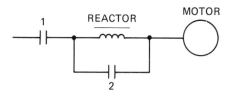

In reactance starting, reactors (coils of wire) are connected in the motor circuit during the starting period and removed for run. This method has essentially the same effect as resistance starting. While operation is more economical than primary resistance starting, the initial installation cost is greater.

Autotransformer Starting

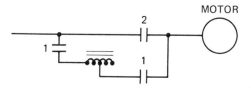

During the starting period the motor is connected to the taps of an autotransformer, thus lowering the voltage applied to the motor terminals.

Star–Delta Starting

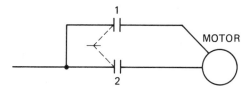

The stator of the motor is STAR connected for starting and DELTA connected for run. To apply star–delta starting, the motor must be wound so that it will run with its stator windings connected in delta. Also, the motor leads must be available at the motor terminal box for field connections.

Part Winding Starting

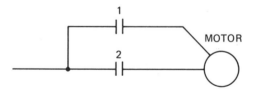

In part winding starting, the stator windings of the motor are made up of two or more circuits. The starter circuit connects the motor winding in series for starting and in parallel for run.

Of all the above methods, primary resistance starting is the most common. In this method a resistor is connected in each motor line to produce a voltage drop that results from the motor current flowing through the resistor. It follows that the voltage applied to the motor is reduced by the amount of voltage dropped on the resistor. As current lowers during the starting period, the voltage available at the motor terminals gradually increases. The result is a smooth and gradual increase in motor torque. At some point during the starting period, a contactor must close to provide a current path around the resistor. At that point full voltage will be applied to the motor terminals. A typical circuit for a primary resistance starter is shown in Figure 4.11.

When the START button of Figure 4.11 is pressed, a circuit is completed from X1 through the STOP button, START button, M coil, the overload contacts, and the fuse to X2. When the M coil is energized, the main contacts M close and the auxiliary contacts M close to provide a maintaining circuit. Motor current must pass through the primary resistors, causing a voltage drop that results in less voltage being applied to the motor. Thus, the motor starts on reduced voltage. After a preset time delay, dependent on the setting of M time closed contacts, a circuit is completed through the S coil which will close the S contacts. The resistors are shunted out and the motor is connected across the lines. If a time closed contact is not available on the motor starter, a separate timing relay may be used. In such cases, the

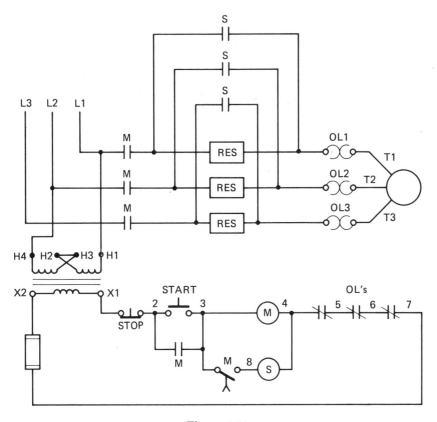

Figure 4.11

coil of the timing relay is connected in parallel with the M coil.

While the relationship of speed and current of a primary-resistance-reduced voltage from starting to full speed depends largely on the driven load on the motor, Figure 4.12 illustrates a typical curve with a moderate load on the motor.

The reactor type of starting is similar to primary resistance, different only in that, in reactor starting, reactors are used in place of resistors. Reactors are large coils of wire and oppose a change in current because of their inductive properties. Reactor starting is much more efficient than resistance starting.

Autotransformer starting is one of the most effective methods of reduced voltage starting. It is not necessary to understand

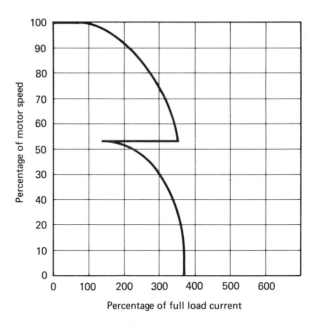

Figure 4.12

fully the electrical characteristics of this method of starting in order to understand the operation of the control circuit. The voltage applied to the motor terminals will remain fairly constant during the starting period while the current drawn by the motor decreases. The decrease in current is due to the motor characteristics.

In autotransformer starting, the starting contactor must have five poles, three of which are placed in the main lines while the remaining two are used to open and close the autotransformer circuit. Figure 4.13 illustrates the start circuit.

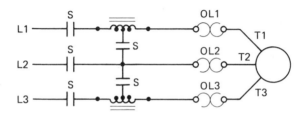

Figure 4.13

The S contacts are closed for the starting period. Once a sufficient amount of time has elapsed, the motor gains speed and the S contacts are opened. Immediately following the starting period, another contactor, with three poles, will close its contacts. This contactor is normally called the RUN contactor, or just R. The power and control circuits are shown in Figure 4.14.

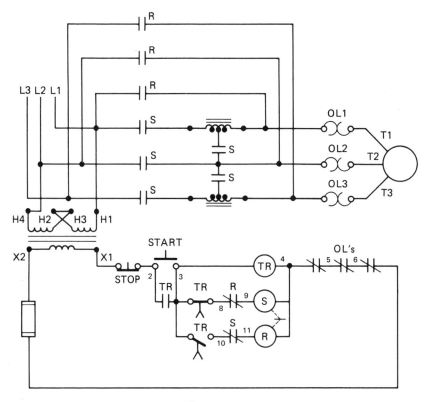

Figure 4.14

The operation of the control circuit is as follows: When the START button is pressed, the TR coil is energized and maintains the circuit through the normally open instantaneous contact TR. The S coil is energized through the STOP button, TR instantaneous contacts, TR time delay contacts, the R interlock contacts, the S coil, the overloads, and the fuse. The S coil closes all S main contacts and starts the motor on reduced voltage. The run coil, R, cannot be energized at this time because of both mechanical and electrical interlocks. After a predetermined time delay, TR time delay contacts operate. The opening of TR contacts in series with the S coil will deenergize the S coil. All S contacts will open. The closing of normally open TR time delay contacts and the reclosing of normally closed S contacts complete the circuit to the R coil. The R contacts will close and connect the motor directly to the line.

To disconnect the motor momentarily from the line is called *open transition*. When the motor is not disconnected from the line when switching from START to RUN, as was the case in primary resistance starting, the condition is called *closed-transition starting*.

Part winding starting is less expensive than most other starting methods because it has no voltage-reducing elements, such as resistors, reactors, or transformers. It uses only two starters and these can be downsized because of the current levels. Overloads should be proportionally smaller for obvious reasons. Transition is closed circuit. The motor manufacturer's specifications should be consulted before applying this method because not all motors can be started by part winding. The circuit for a part winding starter is shown in Figure 4.15.

The control circuit for the part winding starter of Figure 4.15 operates as follows: Pressing the START button energizes the starter coil M1 and timing relay coil TR1. All M1 contacts operate. The main contacts in the power circuit connect L1, L2, and L3 to motor terminals T1, T2, and T3, respectively. The motor starts at reduced current and torque because only half of the motor windings are connected. The M1 auxiliary contacts close to provide a maintaining circuit. After a time delay provided by the TR1 relay, the M2 coil energizes and operates its contacts. The closing of M2 contacts connects L1, L2, and L3 to motor terminals T7, T8, and T9, respectively, applying voltage to the remaining motor windings. The motor is now connected to the line for full current and torque.

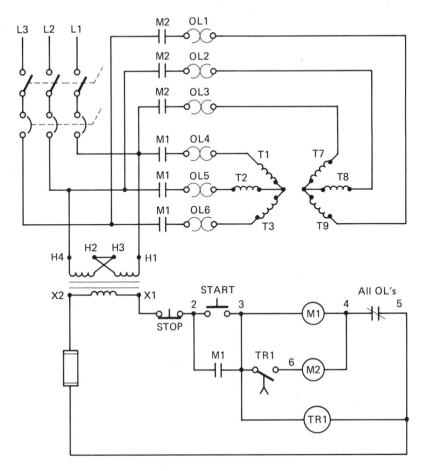

Figure 4.15

Note that the circuit has complete overload protection for each set of motor windings. Again, for part winding starters it is necessary to downsize overload relays because of the circuit configuration.

Chapter 5 Three-Phase Motors

THE AC THREE-PHASE SQUIRREL CAGE MOTOR has appropriately been called the workhorse of industry. The vast majority of motors in use are the squirrel cage type. The motor gets its name from the fact that its rotor construction resembles a squirrel cage.

The three-phase squirrel cage motor operates on the principle of a rotating magnetic field which pulls with it the rotating member of the machine. The magnetic field is set up by the continually increasing and decreasing amount of current flow. This and the arrangement of stator windings produce a rotating magnetic field. Figure 5.1 shows the waveform of three-phase voltages for two complete cycles. The numbers along the base coincide with the diagram below. The dotted lines on the stator diagrams show the condition of the magnetic field for each instant of time indicated on the sine curves. Observe in position 0 that the current is entering the stator on line 1 and dividing equally and leaving through lines 2 and 3. In position 1, current in line 2 is zero and current enters on line 1 and leaves on line 3. Note that the magnetic field has shifted slightly in the clockwise direction. By observing current direction in each phase in each diagram, you should be able to understand how the changes in magnitude and direction of the three-phase currents cause the magnetic field to rotate.

The rotating member of a squirrel cage motor is called a *rotor* and is shown in Figure 5.2. The rotor is made up of copper bars mounted on a core and connected together at the ends with rings. The rotor is mounted in the stator on bearings to permit extreme rotation. The flux of the rotating magnetic field passes through the rotor bars, inducing a current and a corresponding magnetic field in the rotor. The interaction of the two magnetic fields produces a torque, thus causing rotation. As the rotor gains speed, it approaches the speed of the rotating magnetic field. It should be obvious that the speed of the rotor cannot reach the speed of the rotating magnetic field or there would be no induction and no torque. This difference in speeds is called *slip*.

66

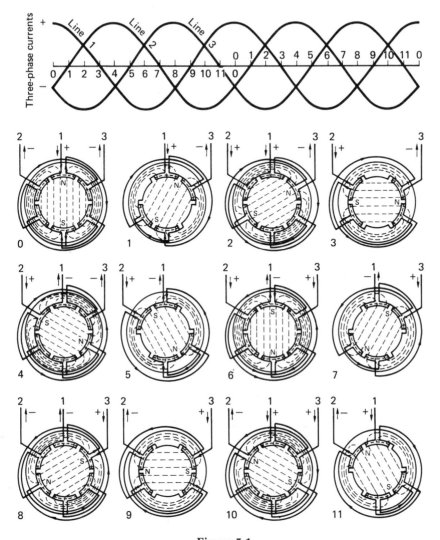

Figure 5.1

The greater the load the greater the slip will be; that is, the slower the motor will run. Even at full load the slip does not increase above no load by a great amount. The motor is essentially a constant-speed motor.

Squirrel cage motors are designed for a wide variety of specific applications and have been classified according to their starting torques and starting currents. Motors with the same

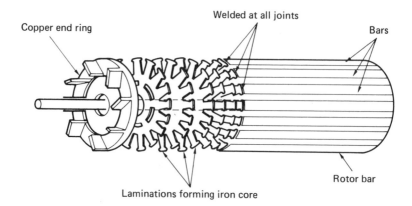

Copper end ring

Welded at all joints

Bars

Rotor bar

Laminations forming iron core

Figure 5.2. Squirrel cage rotor.

Nameplate	Description	Connection
A	Dual voltage	Three-phase, start or wye
N	Dual voltage	Three-phase, delta
F	Dual voltage	Single-phase

Type	Electrical Characteristics
ML	Normal starting torque, normal starting current, normal slip
HML	High starting torque, low starting current, normal slip
WML	Wound rotor, slip ring
MLS	Capacitor start, induction run, single-phase
RLS	Repulsion-induction start and run, single-phase

Type	Mechanical Characteristics
ML	Dripproof
MLU	Splashproof
MLE	Totally enclosed nonventilated
MLF	Totally enclosed fan cooled
MLV	Vertical solid shaft
MLVH	Vertical hollow shaft
GML	Helical parallel shaft motor reducer
GMLW	Right angle worm gear motor reducer
Fluid	Fluid shaft motors

Figure 5.3. Sample nameplates and connection diagrams.

rating, design, and code letters have similar operating characteristics no matter who manufactures them. Figure 5.3 lists sample
nameplates and connection diagrams, electrical characteristics,
and mechanical characteristics.

Motor Terminal Connections

Most three-phase squirrel cage motors have nine leads in the
motor terminal box. Such motors may be connected to high or
low voltage depending on motor design.

The terminal connection is shown on motor nameplates and
for star connected motors should be connected as shown in
Figure 5.4. Occasionally, a motor nameplate is inaccessible,
destroyed, or lost and in such cases, the method shown in Figure
5.5 for determining terminal connection may be used for star
connected motors.

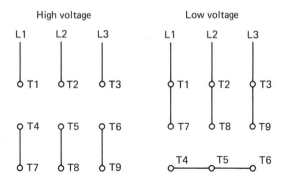

Figure 5.4

For a delta connected motor, the motor terminal connection
pattern is shown in Figure 5.6.

Three-Phase No Tag Test

Occasionally, it is necessary to identify unmarked leads of a
three-phase motor. The method of performing the three-phase no

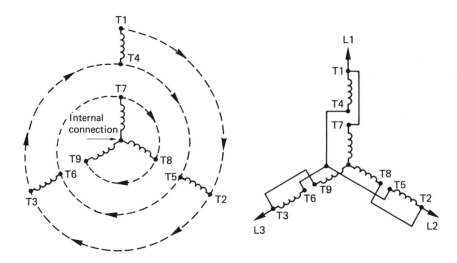

Figure 5.5

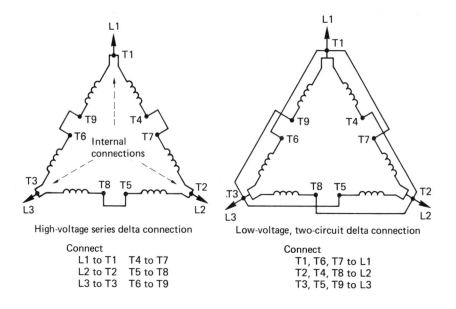

High-voltage series delta connection

Connect
L1 to T1   T4 to T7
L2 to T2   T5 to T8
L3 to T3   T6 to T9

Low-voltage, two-circuit delta connection

Connect
T1, T6, T7 to L1
T2, T4, T8 to L2
T3, T5, T9 to L3

Figure 5.6

tag test requires only a low-voltage transformer, an ac ammeter, and an ac voltmeter and applies only to the standard NEMA lead markings for a Y connected three-phase machine as shown in Figure 5.7.

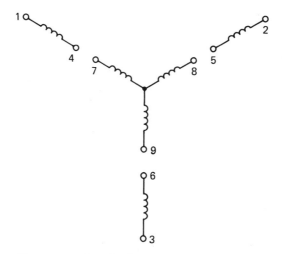

Figure 5.7. Standard marking of motor leads.

Step 1. Locate by any continuity test the three leads from the coils permanently Y connected. See Figure 5.7. Place tags on these three leads and mark them C, D, and E.

Step 2. Locate by any continuity test the pairs of leads from the other three coils. See Figure 5.7. Place tags on one pair of these leads and mark them A and B.

Step 3. Provide a source of 5–10 volts ac and tag the leads W and X. Connect X to coil lead A and W to coil lead B. (These coils have a low impedance and therefore a higher voltage would draw too much current.) See Figure 5.8.

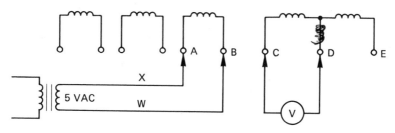

Figure 5.8

Step 4. Connect a low-reading voltmeter (0–10 volts will do) to Y leads C and D. See Figure 5.8. Make a note of the meter readings.

Step 5. Remove the voltmeter lead from Y lead D and connect
it to the third Y lead, marked E. See Figure 5.9. Note the
meter reading

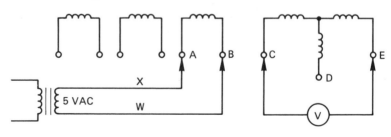

Figure 5.9

Step 6. Remove the other voltmeter lead from Y lead C and
connect it to Y lead D. See Figure 5.10. Note the meter
readings.

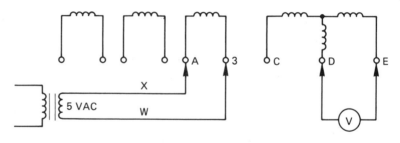

Figure 5.10

There will be a reading on the voltmeter for two of the tests in
steps 4, 5, and 6, but no reading, or 0, for the third test. One Y
lead will be common to these two readings.

EXAMPLE 5.1   Voltmeter connected

                       to C and D reads   2.5 volts
                       to C and E reads   0 volt
                       to D and E reads   2.5 volts
                       then D is the lead common to the readings

D is the Y lead that belongs with the coil A–B to which the ac
supply is connected in the tests. Mark this lead T7. The pair of

coil leads (A and B) used for the test are to be numbered T1 and
T4. The next step is to find which of these leads is T1. In other
words, polarize the coil.

Step 7. Disconnect ac lead W from coil lead B. Disconnect the
voltmeter led from the Y lead you have just marked T7 and
connect an ammeter lead to ac lead W. Connect coil lead B
to T7. See Figure 5.11. Read the ammeter.

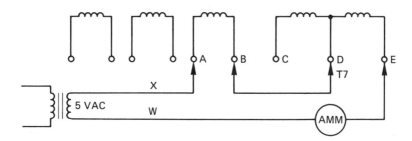

Figure 5.11

Step 8. Remove coil lead B from Y lead T7 and ac lead X from
coil lead A. Connect coil lead A to Y lead T7. Connect
ac lead X to coil lead B. See Figure 5.12. Read the
ammeter.

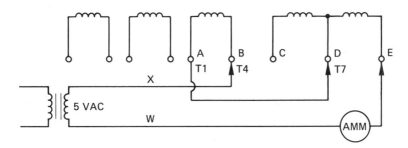

Figure 5.12

One reading will be lower than the other in steps 7 and 8. This is
because in one case the field of coil AB is aiding the field of the Y
coils and so increasing the impedance of the combination,
thereby reducing the current flow. In the other case, the field of

coil AB is bucking the field of the Y coils and so reducing the combined impedance, thereby increasing the current flow.

If the ammeter readings in steps 7 and 8 are the same, you have made a mistake in step 6 and used the wrong Y lead for T7.

The lead of coil AB that was connected to the Y lead T7 when the reading is lowest is to be numbered T4 and the other lead of this pair is T1.

Step 9. Repeat steps 3–8 using another pair of coil leads for A and B and locate another Y lead in step 6. This Y lead can be numbered T8. In step 8 the coil lead connected to T8 when the reading is lowest can be numbered T5 and the other coil lead T2.

Step 10. Repeat the polarization tests of steps 7 and 8 for the remaining pair of coil leads and the third Y lead. The Y lead will be T9. The coil lead connected to T9 when the reading is lowest will be T6 and the other coil lead will be T3.

Note: All leads should now be numbered to correspond with Figure 5.7.

The motor can now be connected for either high- or low-voltage operation. For high-voltage operation, connect L1 to T1, L2 to T2, L3 to T3. Connect T4 to T7, T5 to T8, and T6 to T9.

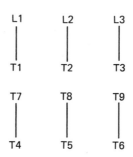

For low-voltage operation, connect L1 to T1 and T7, L2 to T2 and T8, and L3 to T3 and T9. Connect T4 to T5 and T6.

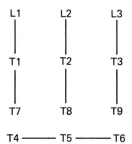

Chapter 6 · Direct Current Machines

CONVERSION OF ONE FORM OF ENERGY into another enables us to use natural power sources as well as manufactured power sources to produce our electrical power supply. Although electricity can be produced by friction, pressure, heat, light, chemical action, and magnetism, the most common method used by large power producers is magnetism.

Numerous dams have been constructed to hold back large amounts of water, providing potential energy that is eventually converted into electrical energy by funneling water through ducts to turn large turbines to drive generators. Steam turbines, diesel engines, gasoline engines, and almost every form of mechanical energy are used today to propel generators to supply the ever increasing demand for electrical power.

Since the most important method by which electricity is produced is magnetism, let us focus on that process. To generate an electrical power source it is necessary to establish a relative motion between a magnetic field and a conductor. The mechanical energy input causes a rotation which results in an output of electrical energy.

Since a basic understanding of magnetism and of the intrinsic characteristics of the metals used for pole pieces is essential to comprehend the functioning of a generator, a magnetic circuit is constructed and analyzed. A current-carrying conductor, shaped to form a coil, produces a magnetized field. If the current is direct in nature, the polarity of the magnetic field is constant and the strength of the field depends on the amount of current and the permeability of air. Permeability is a measure of a material's ability to provide a path for magnetic lines of force. The resultant magnetic field is shown in Figure 6.1.

To improve the efficiency of the magnetic circuit, the air core may be replaced with a metal core. In this case, the nature of the magnetic circuit depends largely on the reluctance of the core metal. A hysteresis loop can be plotted to determine the characteristics of the core metal, since the relationship of the flux

Current

ϕ

Magnetic lines ϕ

North
pole

South
pole

Figure 6.1. Air-core electromagnet.

density and the permeability is nonlinear. In the initial stages
of producing a magnet, or when the coil current is low, the
permeability of annealed sheet steel, for example, is extremely
high, but as the metal begins to "saturate" the permeability
decreases until at saturation it approaches that of air. Saturation
occurs when the magnet reaches a point of strength beyond
which it is not practical to go.

The flux of the magnetic circuit can be varied by varying the
current in the coil, but since the relationship is nonlinear, we
can better see the relationship by plotting the function of the
field current versus the flux. It should be noted that a portion of
the curve is relatively linear. It is a common practice to design
the magnetic circuit of a generator to operate just above the linear
portion of the curve. It is here, at the point referred to as the knee
of the curve, that the most practical control region exists. See
Figure 6.2.

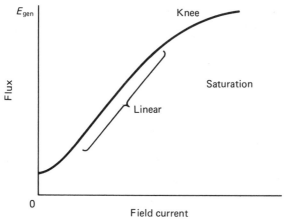

Figure 6.2

Now that we have been introduced to the magnetic circuit, let us examine the effect of flux on the generated voltage. If the speed at which the machine rotates is held constant and only the field current is varied, the generated voltage of the machine follows the curve of the field current versus the flux; that is, the generated voltage follows the saturation curve. The actual value of a machine may be determined by the following equation:

$$E_{gen} = \frac{PZ\,\emptyset\,N}{10^8\,b\,\times\,60}$$

where E_{gen} = the generated voltage
P = the number of poles in the machine
$\emptyset$ = the number of flux lines in the area of the pole piece
N = rpm
b = the parallel paths of the armature conductors

For a given machine, the generated voltage equation may be resolved to

$$E_{gen} = k\,\emptyset N$$

where E_{gen}, $\emptyset$, and N represent the same factors as before and k represents the resultant of the remaining factors, k is constant in a given machine.

From the above explanation, we readily see that the flux and speed alone determine the value of the output voltage of a generator.

It should be obvious now that as the current in the field coil varies, the generator output also varies. If the machine is operating at the knee of the saturation curve, it is producing as much flux as is practical for the amount of input current. Thus, the generator is operating in the most efficient region.

The speed of the machine, more precisely, is the speed of the coils of wire that are arranged in such a manner as to permit extreme rotation within the magnetic field. The coils of wire, called the *armature*, are wound to provide the most efficient and practical circuit possible. We can simplify the armature circuit by condensing it to a single coil of wire with provisions made for a sliding connection to maintain contact with a stationary wire to accept the output voltage from the armature coil. Figure 6.3 illustrates the machine described thus far. It must be realized that the sketch represents a simplified model and not the actual machine construction.

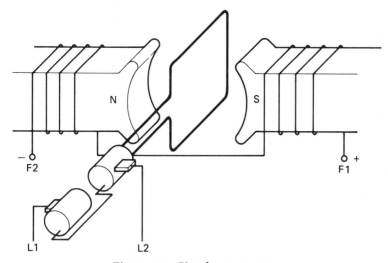

Figure 6.3. Simple generator.

Since the value of the generated voltage depends on the number of flux lines cut in a given period of time, as the armature rotates the number of lines cut varies as the angle of the interception varies. Thus, voltage values for the complete rotation vary as the sine of the angle. From this we can establish that the value of the voltage for a given position of the armature coil is a function of the sine of the angle. We can now write the equation for the instantaneous voltage values:

$$e = (E_{max}) \sin L$$

where e = the instantaneous voltage value

E_{max} = the maximum generated voltage, or the voltage generated at 90° (this corresponds to E_{gen} previously mentioned)

$\sin L$ = the sine function of the instantaneous angle

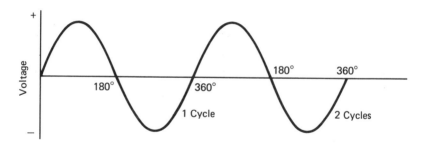

Figure 6.4. Sine curve.

By plotting the instantaneous values of the generated voltage, we can generate the sine curve shown in Figure 6.4.

As can be seen from the sine curve, the voltage changes in polarity after the machine has rotated through 180°. Instead of a strict back and forth motion, the conductor is rotating through the field and in so doing, the general direction of travel reverses each 180°. For this reason, the polarity, or direction, of the voltage reverses.

In summary, to generate a voltage, mechanical energy produces rotation of an electrical conductor through a magnetic field. The value of the generated voltage depends on the number of flux lines cut in a given time period. For a given machine, the voltage generated depends on the strength of the field and the speed of the machine.

Electric Generators and Motors

Electric generators and electric motors are dynamos that convert mechanical energy into electrical energy and electrical energy into mechanical energy. A dynamo consists of two basic parts—the stationary part and the rotating part. The most important member of the stationary part is the field. The frame, which houses the field windings and their laminated pole pieces, also contains the brush rigging and the bearings for the rotating member.

The rotating member is made up of a laminated steel core, with insulated copper armature windings embedded in slots.

Extreme rotation is permitted by mounting the rotating member on a shaft which is supported by bearings of the stationary member. The coils of the armature windings are fastened to the commutator which is positioned so that the brushes held by the brush rigging may rest on it. The brushes are held against the commutator by spring tension to provide a good sliding connection between the armature coils and the external circuit.

Direct Current Generator

In studying the characteristics of a dc generator, we use many of the laws of electric and magnetic circuits. It follows then that an understanding of the law of electromagnetic induction is a basic requirement for understanding the operation of a dc generator

Whenever an electric current flows in a conductor, there is a corresponding magnetic field. Although the presence of a magnetic field does not necessarily mean that there must be a current in the conductor, it sets up the possibility of inducing a current in the conductor. By establishing a relative motion between the conductor and the magnetic field an EMF is produced.

When a conductor cuts magnetic lines of force, an electromotive force is set up. The electric generator operates on this fundamental rule. This principle is illustrated in Figure 6.5. In summary, the principle of generator action requires the presence of a magnetic field and a moving conductor to cut the flux.

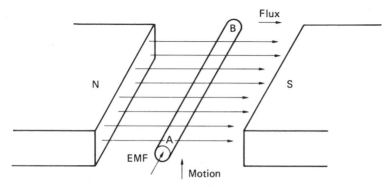

Figure 6.5. An electromotive force is induced in a conductor passing through a magnetic field.

Right-Hand Rule

It should be observed in Figure 6.6 that three directions are established.

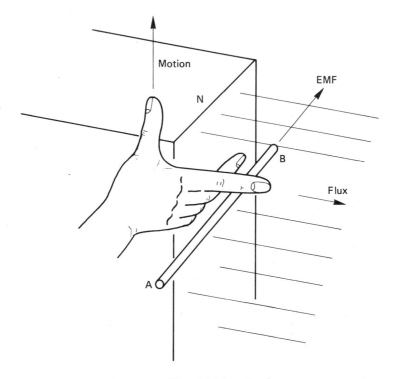

Figure 6.6. The right-hand rule.

1. The conductor is passing up through the magnetic field.
2. The magnetic lines of force are from the north pole (N) of the magnet to the south poles (S).
3. The direction of EMF is from point A to point B—into the page.

Reversing either the direction of motion or the polarity of the magnets would result in a change in the direction of the induced EMF.

Right-Hand Rule: Extend the thumb, forefinger, and middle finger of the right hand at right angles to each other. Place the hand so that the *thumb* points in the direction of the motion and the *forefinger* in the direction of the flux. The *middle finger* will point in the direction of the induced electromotive force as shown in Figure 6.6.

Voltage Values: Faraday's Law

The magnitude of the generated voltage is directly proportional to the rate at which a conductor cuts magnetic lines of force. If a conductor cuts 10^8 lines of flux in 1 second, 1 volt is induced in that conductor. Stated algebraically,

$$E = \frac{\emptyset}{t \times 10^8}$$

where E = generated voltage in a conductor
$\emptyset$ = total flux cut
t = time in seconds

EXAMPLE 6.1

A conductor moves at a rate of 30 times a second through a magnetic field whose flux density is 20,000 lines per square inch. The pole face of the magnet measures 20 inches by 40 inches. What is the induced EMF in the wire?

Solution:

Total flux = flux density B times the area of the pole face.
$\emptyset = BA$
where $\emptyset$ = total flux
B = flux density (lines per square inch)
A = area of the pole face

Then,

$$\emptyset = BA = 20 \times 10^3 \ \frac{lines}{in.^2} \times 20 \ in. \times 40 \ in.$$

$$\emptyset = 16 \times 10^6 \ lines$$

and

$$E = \frac{\emptyset}{t \times 10^8} = \frac{16 \times 10^6 \text{ lines}}{(1/30 \text{ sec}) \times 10^8}$$

$E = 4.8$ volts

Direct Current Motor Principles

An electric motor converts electrical energy into mechanical energy. A generator converts mechanical energy into electrical energy. The structure of the two machines is essentially the same so the same machine may be operated as a motor or as a generator. In fact, it is important to remember that when the machine is acting as a motor there is a generator action in the motor, and when the machine is acting as a generator there is a motor action in the generator.

The operation of a dc motor depends on the principle that a conductor carrying current in a magnetic field tends to move at right angles to the field. To understand the operation of a motor, we must study the arrangement of the magnetic fields and how they are produced and controlled. We must also understand the *current-carrying conductors* or armature windings, commutator, and brushes.

Since the operation of the motor depends on magnetism, we must know the following rules:

1. If the conductor is grasped in the right hand with the thumb pointing in the direction of the current, the fingers will point in the direction of the magnetic field.
2. Grasp the coil with the right hand so that the fingers point in the direction of the current in the coil. Then the thumb extended longitudinally will point in the direction of the flux, or toward the north pole.

With these rules we can determine the effects on the current in the motor. The sketch of Figure 6.7 shows the magnetic field between the opposite poles of two magnets produced by coils of

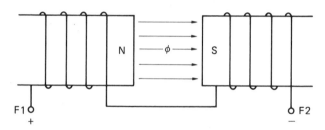

Figure 6.7

wire. These magnets must be controlled in order to control the motor. The coils must be wound in such a manner as to provide the proper relationship of the coils; that is, by controlling the direction of current in the coils we can control the polarity of the magnets. Then by controlling the amount of current in the coil we can control the strength of the magnets.

The current through the armature must also be controlled. Since the armature is the rotating member, a sliding connection must be provided. Also, the direction of current in the armature coils must be reversed each half revolution so that a constant direction of torque may be produced. The commutator provides the sliding connection, and thus serving as a reversing switch. In a generator where the output is electrical, the commutator brings to the brush surface the voltage of each armature coil in the circuit. These voltages add up to present the total operating voltage of the generator at the brushes of the machine. The function of the brushes is to carry the current to and from the commutator. The brushes also prevent sparking and coat the commutator with a near friction-free film to reduce wear to both the brushes and the commutator. It is important that the proper brush grade be used.

In Figure 6.8 we are using principles to show the forces on the armature of a two-pole motor. The current is flowing from A1 to A2 and also from F1 to F2. When current is caused to flow in the armature, there is an interaction of the flux produced by the field windings and the flux produced by the armature windings. Careful analysis of the magnetic forces will show the resultant torque.

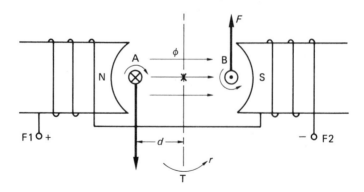

Figure 6.8

The magnetic field around conductor A causes a distortion of the flux field so that the field above conductor A is strengthened and the field below conductor A is reduced. Thus, there is a downward force F exerted on conductor A. Note that the reverse conditions are true on conductor B, causing an upward force to be exerted on conductor B. It can further be noted that as these conductors travel through the neutral zone (parallel to the flux lines produced by the field windings) the current direction in the armature coil must be reversed so that conductor A can exert an upward force during that half of the revolution that it has to travel in the upward direction. The function of the commutator is to provide this current reversal in the armature coils. Figure 6.9 illustrates this.

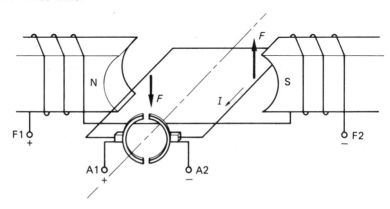

Figure 6.9

The coming current to the armature is determined by the line voltage and the resistance of the armature circuit. As the armature begins to turn and the armature conductors cut through the magnetic lines of flux, a voltage is generated in the armature that is proportional to the speed of the armature and the strength of the field:

$$I_A = \frac{E_A - E_{gen}}{R_A + R_{stg}}$$

The motor speed can now be controlled by the generated voltage, or the amount of resistance in series with the armature.

Machine Components and Symbols

Armature

The armature is the rotating member. Its coils are connected to the commutator for connection to the external circuit.

Shunt Field

The shunt field is constructed of many turns of fine wire wound around a pole piece. Its function is to produce a magnetic field whose strength is independent of the load on the machine. It is usually connected across the line. The magnetic field strength is varied by placing an adjustable resistor, or rheostat, in series with the shunt field windings. The current in the field windings is a factor in the output of a generator and the speed of a motor. The rheostat is connected so that a clockwise rotation increases the function.

Series Field

$$\underset{\text{SERF}}{\underline{\text{S1} \quad \text{—}\!\text{mm}\!\text{—} \quad \text{S2}}}$$

The series field is constructed of a few turns of heavy wire wound around a pole piece. Its function is to produce a magnetic field whose strength is a function of the load on the machine. The series field is connected in series with the armature so that as load increases field strength also increases.

Motor Types

dc Shunt Motor

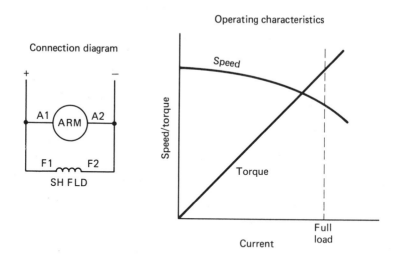

Shunt motor speed varies slightly from no load to full load. It has a definite no load speed. Its speed can be controlled above base speed by weakening the field, and below base speed by lowering the applied voltage. Base speed is the speed at which the motor will run with rated armature voltage and field current applied.

dc Series Motor

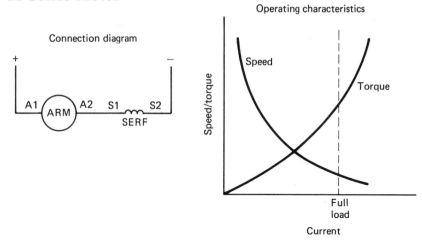

Series motor speed varies greatly as load changes. Care must be taken to be certain that the motor is not operated without a significant load to prevent overspeeding. As load is reduced and armature current lowers, the field strength lowers and causes an increase in speed. The motor will self-destruct at no load. The torque produced increases with the square of the armature current.

dc Compound Motor

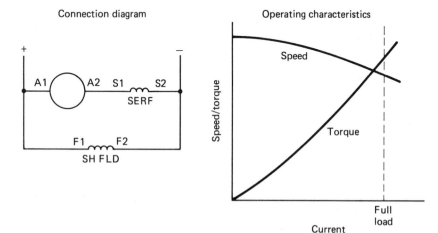

The compound motor contains both a shunt field and a series field and therefore has characteristics between the shunt and the series motors. The motor has the good starting torque characteristics provided by the series field, while the shunt field provides for a relatively constant speed. The characteristics curves show that the speed variation from no load to full load is less on the compound motor than on either the shunt or series motor.

Figure 6.10 provides approximations of full-load currents derived from average values for representative motors of their class.

	Full-Load dc Motor Current at		
Horsepower	115 volts	230 volts	550 volts
0.25	—	—	—
0.50	4.5	2.3	—
0.75	6.5	3.3	1.4
1	8.4	4.2	1.7
1.5	12.5	6.3	2.6
2	16.1	8.3	3.4
3	23	12.3	5.0
5	40	19.8	8.2
7.5	58	28.7	12
10	75	38	16
15	112	56	23
20	140	74	30
25	185	92	38
30	220	110	45
40	294	146	61
50	364	180	75
60	436	215	90
75	540	268	111
100	—	357	146
125	—	443	184
150	—	—	220
175	—	—	—
200	—	—	295

Figure 6.10

Chapter 7 Direct Current Motor Control

DIRECT CURRENT MOTORS above fractional horsepower require series resistance in the armature circuit during the starting period. The starting resistance is removed in steps as the motor gains speed and generates counter EMF. The maximum permissible starting current depends on the size of the motor and the application. Starting current is usually limited to about 200% of full-load current. The value of the generated voltage is proportional to the speed of the armature and the strength of the magnetic field. Stated algebraically, $E_{gen} = k\,\emptyset \times \text{rpm}$.

Armature current produces the torque that causes rotation and is determined by three factors:

1. The applied voltage, E_a.
2. The counter EMF, E_{gen}.
3. The armature resistance, R_A

During the starting period, the armature increases in speed and the current lowers. This is because the generated voltage buildup reduces the effective voltage across the armature. As current lowers the torque, the armature acceleration also lowers. When the torque produced equals the torque provided by the load, the armature will stop accelerating and will run at a constant speed. The armature will draw current from the line to produce a torque equal to the countertorque. The torque/countertorque relationship determines the speed condition of the armature; that is, whether speed will increase, decrease, or remain constant. In other words, the speed of an armature will "regulate" until the torque produced is equal to the amount of load placed on it. Three conditions may exist:

1. Torque is equal to countertorque—speed remains constant.
2. Torque is greater than countertorque—speed increases.
3. Torque is less than countertorque—speed decreases.

Armature speeds below base speed may be controlled by varying the voltage applied to the armature and above base speed by varying the strength of the field, thus making the machine's speed controllable over an extremely wide range.

Lowering applied voltage obviously lowers current and torque. Reducing field current weakens the field, causing a reduction in generated voltage. As a result, armature current and torque will increase. The field current and field flux relationship may be seen in Chapter 6, Figure 6.2.

Manual Motor Starter

A manual motor consists of a *starting box* with resistance connected in the armature circuit and gradually removed by manually advancing a control lever. In small-motor starters, the resistance is usually made of coils of high-resistance alloy wound on tubes of heat-resistant insulating material. Large-motor starters use resistors made of iron grids. The controller consists of an arm arranged to pivot at one end with the other end arranged to slide over a set of contacts also mounted in an insulating material. The resistance coils are connected to the contacts. The circuit is determined by the position of the controller. As it is advanced from the OFF position to FULL SPEED position, the starting resistance is gradually removed from the armature circuit.

When the controller reaches the FULL SPEED position, it is held there by an electromagnet whose coil is connected in the field circuit. If the current supply is cut off because of a blown fuse or other power failure, the controller returns to the OFF position by a spring mounted in the controller. This arrangement is called *low-voltage release* and prevents a sudden across-the-line start once power is resumed. Figure 7.1 shows the circuit for a manual motor starting box.

A plot of armature current during the starting period is shown in Figure 7.2. As each portion of resistance is removed from the armature circuit, current increases momentarily, resulting in further acceleration. Figure 7.3 shows the relationship of speed

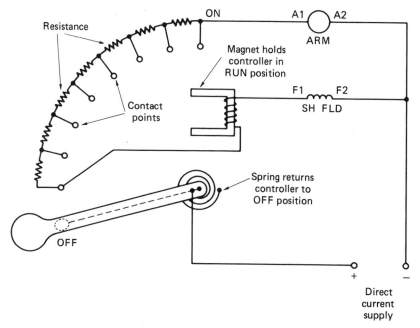

Figure 7.1

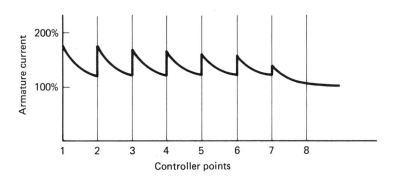

Figure 7.2

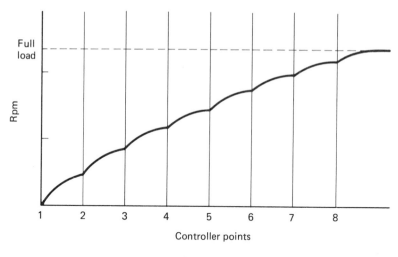

Figure 7.3

when the controller is advanced to position 1 at time 1, position 2 at time 2, and so on.

Magnetic Motor Starter

The principles of a manual motor starter apply to a magnetic motor starter except that the switching is done through the use of magnetic relays and contactors. Magnetic motor starters can be semiautomatic or automatic depending on the needs of the system. A basic two-step semiautomatic dc motor controller is shown in Figure 7.4. The circuit includes a master switch, a magnetic overload relay, a control relay, and two main contactors.

The operation of the circuit of Figure 7.4 is as follows. With the master switch in the OFF position, the normally closed contact in the master switch is closed. With the control knife switch (CKS) closed, current will flow from the positive line, through the control knife switch, the fuse, #3 reset contact, UV coil, the normally closed overload contact, the fuse and control knife switch in the negative line, to energize the UV coil. UV contacts will close. When the master switch is moved to the first

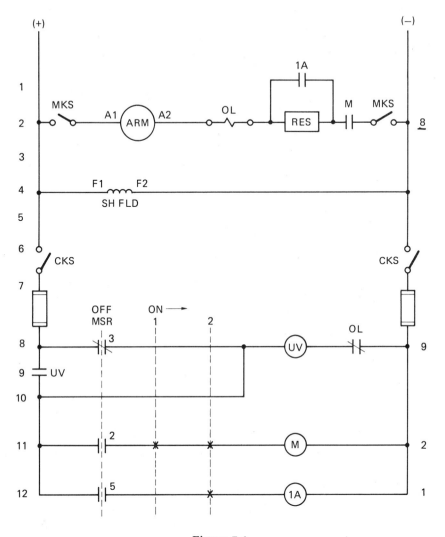

Figure 7.4

ON position, the #3 reset contact opens and the #2 master switch contact closes. The UV coil circuit is maintained through UV contacts. This arrangement provides low-voltage release and prevents a sudden across-the-line start once power is resumed following an overload or power failure. In this situation, the master switch must be returned to the OFF position to energize the UV coil through the reset contact. The closing of the #2 master switch contacts completes a circuit through the M coil, energizing it. The M contacts in the power circuit close to complete the armature circuit and start the motor. After a brief time period, the master switch may be advanced to the second ON position to close the #5 master switch contacts. This completes the circuit through the 1A coil which energizes and closes the 1A contacts around the starting resistance. The motor accelerates to its operating speed. Moving the master switch back to the OFF position opens the control circuit to deenergize the coils. The power circuit contacts will open and the motor will coast to a stop.

Ohm's law is used to determine the starting resistance value. If we want to limit the starting current to 70 amperes when operating on a 240 volt dc supply, with an armature resistance of 0.5 ohms, the resistance needed is

$$R_{stg} = \frac{E}{I} - R_A = \frac{240}{70} - 0.5 = 2.93 \ \Omega$$

Dynamic Braking

Dynamic braking makes use of the generator action in a motor to bring it to rest. The polarity of the counter EMF is opposite to the polarity of the applied voltage. When the motor is disconnected from the line and continues to coast, the counter EMF continues to be generated provided the field remains energized. Obviously, the motor eventually coasts to a stop; the time it takes to stop depends on the inertia of the moving parts and the bearing, brush, and wind friction. However, if the armature terminals are disconnected from the source and immediately connected across a resistor, the motor will stop more rapidly. A braking force is created by the current flow caused by the counter EMF and the newly established resistor circuit. The mechanical energy that is stored in the moving parts is converted to electrical energy and is dissipated across the resistor in the form of heat. When the mechanical energy is expended, the motor comes to rest. If the value of the resistor is made low enough, the energy is dissipated very rapidly and the motor stops quickly.

Since the dynamic braking resistance determines the maximum value of the current in the armature circuit as well as the stopping time of the motor, it is necessary to calculate its ohmic value on the basis of the permissible current at the instant the braking action is initiated. This is because the counter EMF, which is responsible for the current, is a maximum at that instant and diminishes rapidly to zero as the motor comes to a stop. The following example illustrates how the dynamic braking resistance is determined.

EXAMPLE 7.1

For a 10 hp motor operating on a 240 volt supply at full load, calculate the resistance of a dynamic braking resistor if the armature current is to be limited to 1.75 times its rated value. The armature resistance is 0.5 ohm, and full-load armature current is 40 amperes.

Solution:

$$E_G = E_A - (I_{FL}R_A)$$

$$E_G = 240 - (40 \times 0.5)$$

$$E_G = 220 \text{ volts}$$

$$R_{DB} = \frac{E_G}{1.75\ I_{FL}} - R_A$$

$$R_{DB} = \frac{220}{1.75\ (40)}$$

$$R_{DB} = 2.64 \text{ ohms}$$

The circuit of Figure 7.5 illustrates how dynamic braking is usually set up. It is essentially the same as the circuit shown in Figure 7.4 with the addition of the DB circuit. When the motor is shut off, the main contactor (M) is opened to disconnect the

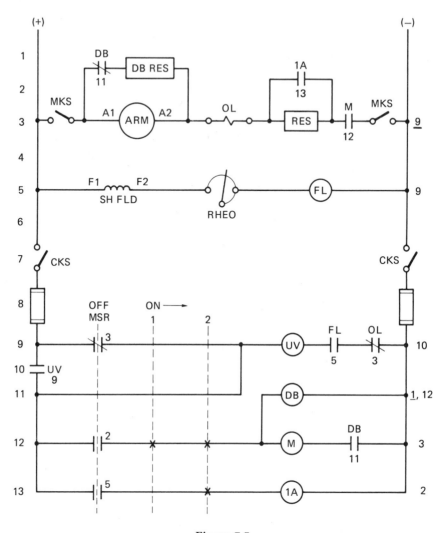

Figure 7.5

motor from the line, and at the same time, the dynamic braking
contactor (DB) is closed to connect the dynamic braking resistor
across the armature. The direction of current flow through the
armature during dynamic braking is reversed to supply the
braking force.

In summary, dynamic braking is accomplished by using the
motor involved as a generator, taking it off the line and applying
an energy-dissipating resistor to the armature. As the motor

slows down, the generator action becomes less, the current becomes less, and the braking becomes less. Therefore, a motor cannot be stopped by dynamic braking alone. Used in conjunction with a friction brake, a motor can be stopped rapidly and held in that position until released.

Reversing Control

Reversing the direction of rotation of a dc motor is achieved by reversing the direction of current through either the armature or the field windings. Because of the inductive kick that is produced when the field circuit is opened, the armature current is usually reversed. This is done by using two pairs of contacts: one set for forward, and one set for reverse. The contacts are coil operated as directed by a reversing master controller. In the circuit of Figure 7.6, when 1F and 2F contacts close for forward operation, current is directed from the positive line through the armature from terminal A1 to A2. For reverse operation, contacts 1R and 2R close and current flows from A2 to A1. Interlocking is provided to ensure that the forward and reverse circuits do not energize simultaneously. Electrical interlock is provided by normally closed 2R and 1F auxiliary contacts. Mechanical interlock is also provided.

Care must be taken when going from forward to reverse to prevent excessive current from flowing. If a reversing circuit is established while the motor continues to coast in the forward direction, the counter EMF and the applied EMF will be of the same polarity and a high current will flow unless resistance is added. Circuits may be designed to provide plugging control or antiplugging control, depending on the needs of the operation.

Contact Sequence Chart

The table shown with the circuit of Figure 7.6 is the contact operating sequence chart. It indicates the contacts that are closed

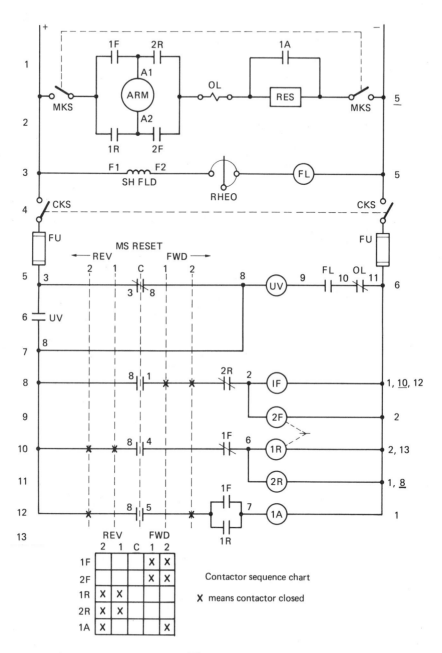

Figure 7.6

at each position of the controller. This chart is a valuable aid in troubleshooting.

Note from the chart that all contacts are open when the controller is in the center position. On first position forward, 1F and 2F are closed, and all other contacts are open. On second position forward, 1F, 2F, and 1A contacts are closed, and all other contacts are open.

Field Rheostat

The circuits of Figures 7.5 and 7.6 include rheostats (variable resistors) in the shunt field circuits. If the rheostat is rotated in the clockwise direction, the resistance of the shunt field circuit is increased. The result is a reduction in field flux which causes a reduction in the counter EMF. At that instant, both armature current and torque increase to cause an increase in speed. As speed increases, both current and torque become progressively less until the speed is stabilized at the point where the torque equals the countertorque.

FL Relay

To ensure that the motor speed does not become excessive because of reduced field current, a FL relay is used. The relay coil is placed in series with the shunt field. The relay contacts are placed in the UV coil circuit. The relay is a special-purpose relay with a specific, and adjustable, drop-out current. That is, a certain minimum current is required in the coil to energize the coil sufficiently to keep the contacts closed. The coil pulls against an adjustable spring. If the spring tension is increased (by manual adjustment), the relay will drop out at a higher current level, thus limiting the speed to the speed that corresponds to that field current. The sketch in Figure 7.7 illustrates the principles of operation of a FL relay. Reducing spring tension permits the relay to hold in at a lower current level and the motor to operate at a higher rpm.

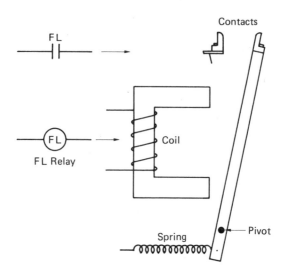

Figure 7.7 The force produced by the coil is opposed by the spring force. If current in the coil falls below the set point, the relay will open.

Time–Current Acceleration Relay

The accelerating time delay for Figure 7.8 is controlled by an EC&M, type AR, acceleration relay. This relay consists of a single series coil, an inductor tube, and a set of normally closed contacts. On normal or light loads, it provides for an accelerating time interval that varies with the load. On unusually heavy starting loads, where the motor may stall, it forces acceleration until either the motor starts or the overload condition turns off the control.

When current passes through the series coil, the magnetic flux build up in the core induces a current in the inductor tube and causes it to pop up and open a set of contacts. The contacts are connected in the next accelerating contactor coil circuit. When the series current in the accelerating relay coil stops increasing, the inductor tube begins to settle back down. If the series current remains high, the closing rate is slower than it would be for a lower current. The total time delay is provided by

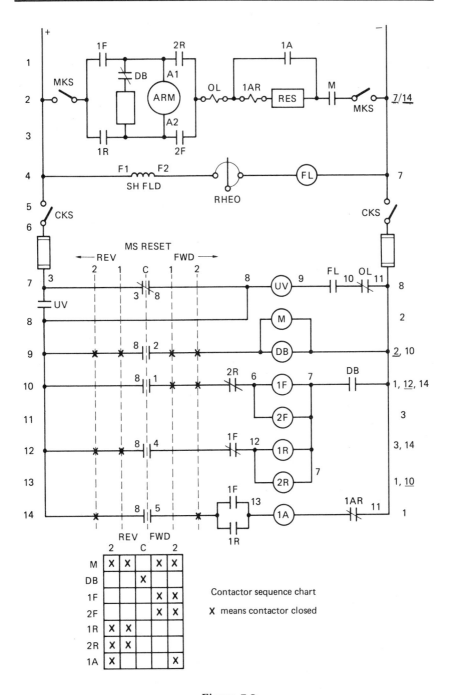

Figure 7.8

the popping action of the inductor tube and the time required for it to settle. At the bottom of its stroke, a set of contacts closes.

An adjustable time delay is provided by an iron core placed inside the inductor tube. The maximum time setting is obtained with the iron core screwed up into the frame flush with the lock nut. Screwing the core down shortens the time interval for a given coil current. See Figure 7.9.

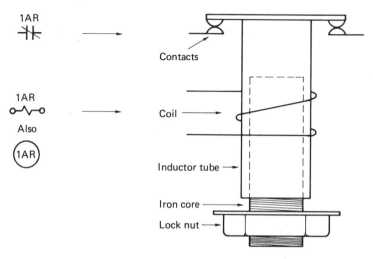

Figure 7.9 To increase the time delay, screw the core higher up into inductor tube.

Plugging Relay

The plugging control for Figure 7.10 is provided by an EC&M, type FKP, plugging relay. The relay has a shunt wound coil and a single set of normally closed contacts. The relay is designed for use on dc reversing/plugging magnetic controllers. It is made polarity sensitive by a small rectifier circuit connected in series with the relay operating coil. The relay has a close differential between pick up and drop out and provides plugging protection above any appreciable speed. The relay tends to force acceleration after plugging if line voltages are below normal and to retard acceleration on voltages above normal.

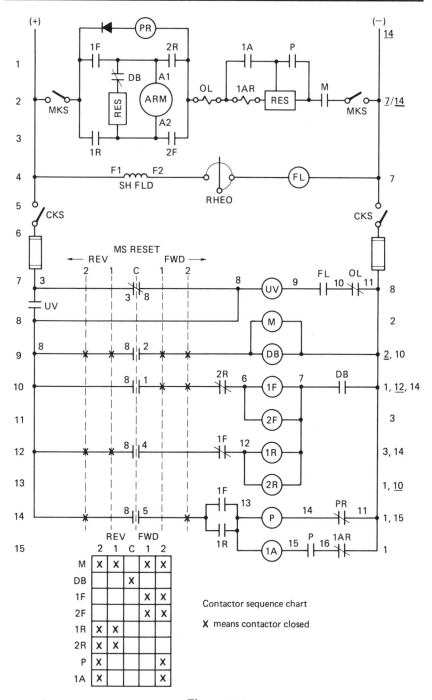

Figure 7.10

When starting from rest, either forward or reverse, the relay coil does not energize because of the blocking effect of the rectifier. After the motor has accelerated to 10–15% of normal speed and then is plugged, a voltage is applied to the armature with the same polarity as the counter EMF. Under this condition, a voltage is applied to the relay coil and the rectifier in the conducting direction and is approximately proportional to the motor's speed at the instant of plugging. The relay then energizes and prevents closure of the plugging contactor and the sequential closing of the accelerating contactor.

It is not only necessary to hold back the operation of the accelerating contactor but resistance must be added to the circuit to prevent excessive current.

Starting-Plugging Resistance Calculation Procedure

The following example is a review of the operating characteristics of a dc starter equipped with reversing and controlled plugging.

EXAMPLE 7.2

For the following circuit and values, find the value of resistance for each step. Assume starting under a plugging condition.

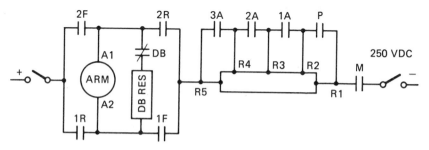

The resistance of the armature is 0.5 ohm. The line voltage is 250 volts dc. The full-load current is 40 amperes. The maximum permissible current is approximately 75 amperes.

Solution:

To limit current to a maximum permissible value, resistance must be added to the circuit. The value of the circuit resistance may be found by the following equation:

$$R_{c1} = \frac{E_{line} + E_{cEMF}}{I_{max}}$$

where

$$R_{c1} = \text{resistance of the circuit on first point}$$
$$E_{line} = \text{the source voltage}$$
$$E_{cEMF} = 0\text{--}250 \text{ volts, depending on speed}$$
$$I_{max} = \text{maximum permissible current} = 75.2$$
$$\text{amperes}$$
$$R_A = 0.5 \text{ ohm} = \text{the armature resistance}$$
$$I_{full\ load} = 40 \text{ amperes}$$

Inserting values into the above equation, we find the resistance of the circuit on the first point. This is the resistance of the armature and the entire starting resistance (R_A + R5 to R1).

$$R_{c1} = \frac{250 + 250}{75.2} = \frac{500}{75.2} = 6.64 \text{ ohms}$$

Assuming the plugging circuit permits the control to advance to the second point when speed is zero, we write the following equation and solve for the resistance in the circuit on second point.

$$R_{c2} = \frac{E_{line} - E_{cEMF}}{I_{max}} = \frac{500}{75.2} = 3.32 \text{ ohms}$$

The high current in the armature causes the motor torque to exceed the countertorque and the motor accelerates, the counter EMF increases, and the current decreases. If the motor is subjected to full load, the current can only decrease to 40 amperes. At that point, we calculate $I \times R$, which is $E_{line} - E_{cEMF}$ and is also equal to the effective voltage:

$$I \times R = 40 \times 3.32 = 133 \text{ volts}$$

Advancing the control to the third point and removing some of the circuit resistance cause the current to increase and cause further acceleration. We wish to limit current to I_{max} so we must again determine the circuit resistance. Since we know $E_{line} - E_{cEMF} = I_{full\ load} \times R_c$, we can write

$$R_{c3} = \frac{E_{line} - E_{cEMF}}{I_{max}}$$

in the following form:

$$R_{c3} = \frac{I_{full\ load} \times R_{c2}}{I_{max}}$$

Inserting values we find

$$R_{c3} = \frac{133}{75.2} = 1.77 \text{ ohms}$$

Again, because of the increased current, the motor accelerates, the counter EMF increases, and the current decreases to full-load current. To cause further acceleration, we must again remove resistance. The value left in the circuit can be determined as before:

$$R_{c4} = \frac{I_{full\ load} \times R_{c3}}{I_{max}} = \frac{40 \times 1.77}{75.2} = 0.941 \text{ ohm}$$

Again, since the current has increased, the motor accelerates, the counter EMF increases, and the current decreases to full-load current. To cause further acceleration, we must again remove resistance. Since this is the last point of control, the only resistance remaining in the circuit is the resistance of the armature. The resistance in the circuit on this last point is then 0.5 ohm: Therefore, $R_{c5} = 0.5$ ohm. It is necessary here to check the incoming current to see that it does not exceed I_{max}. Transposing the previously used equation to solve for I_{max5}, we find

$$I_{max} = \frac{I_{full\ load} \times R_{c4}}{R_{c5}} = \frac{37.6}{0.5} = 75.2 \text{ amperes}$$

Since the incoming current is equal at each point, our resistance values must be correct. It is now necessary to determine the values removed by each shunting contactor.

At the first point we had 6.64 ohms in the circuit. At the second point we had 3.32 ohms in the circuit. We removed the difference in the two, $R_{c1} - R_{c2}$, which is the value removed by the first shunting contactor. We follow the same procedure for the remaining values:

$R_{c1} - R_{c2} = 6.64 - 3.32 = 3.32$ ohms = R1 to R2

$R_{c2} - R_{c3} = 3.32 - 1.77 = 1.55$ ohms = R2 to R3

$R_{c3} - R_{c4} = 1.77 - 0.941 = 0.829$ ohm = R3 to R3

$R_{c4} - R_{c5} = 0.941 - 0.5000 = 0.441$ ohm = R4 to R5

Antiplugging

In applications where the driven load cannot sustain a sudden reversal of direction, antiplugging control is used. With antiplugging control the motor must come to a very low speed before it can be restarted. The coil of the antiplugging relay is connected across the armature. When the motor starts and line voltage is applied to the armature, the coil energizes and opens a set of normally closed contacts that are connected in the start circuit. Once the motor is turned off and continues to coast, the counter EMF is applied to the relay coil to keep it energized. The relay is designed so that both pick-up and drop-out voltages can be adjusted. The drop-out voltage corresponds to a certain low speed of the armature. The circuit of Figure 7.11 includes antiplugging control.

Field Accelerating Relay

The field accelerating relay is designed to provide for full field starting for a dc shunt motor without changing the setting of

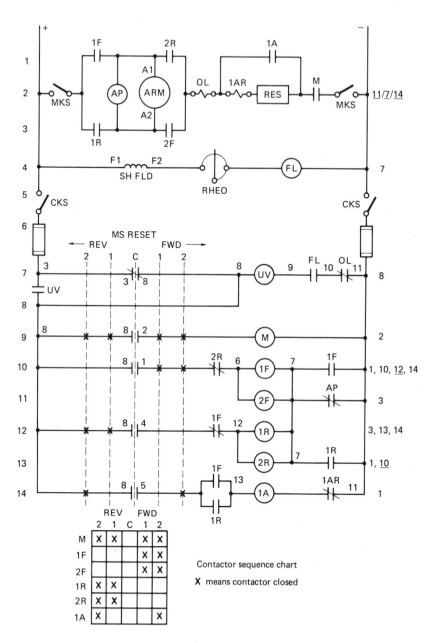

Figure 7.11

the field rheostat. The relay has two coils wound on the same pole piece, either of which is capable of independently closing a set of normally open contacts. The required closing force is adjustable by varying the magnetic air gap and the spring tension. One of the coils is a series coil designed to operate on armature current. The other is designed to operate on line voltage.

The contacts of the relay are connected in parallel with the shunt field rheostat and are closed by the coils during the starting period. On the final accelerating step, the voltage coil is deenergized. If armature current is sufficiently high at this stage of the starting procedure, the contacts will be held closed by the series coil. FA contacts remain closed as the armature current decreases until the current in the series coil is not sufficient to keep the contacts closed. When the contacts open, the field rheostat is placed in the circuit to decrease the shunt field current and reduce the strength of the motor's magnetic field. The reduction in flux produces less counter EMF and thus armature current increases. The increase in armature current may reenergize the relay coil depending on relay adjustments. The procedure continues until the motor has gained full speed.

The FA relay is sometimes called a flutter relay because of its operating characteristics on the final accelerating stage. Figure 7.12 illustrates the components of a FA relay.

A FA relay is used in the control circuit of Figure 7.13.

Inductive Time-Limit Contactors

The accelerating contactors used for inductive time-limit acceleration control have two coils—one for operating the contactor and the other for holding it open. The operating coil operates on line voltage. The holding coil can either be connected parallel to the starting resistors or be placed in the control circuit in series with a resistor to decrease the coil voltage. The coils are arranged so that the magnetic forces are in opposition. The strength of the coils is so proportioned that with full line

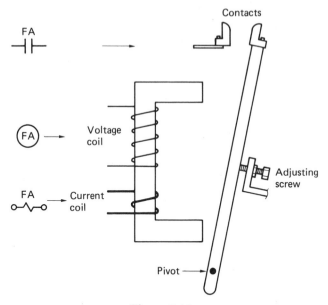

Figure 7.12

voltage on the operating coil, and a very small voltage on the holding coil, the contactor will not operate. The time delay is provided by flux decay after the holding coil is deenergized. The air gap in the magnetic circuit of the closing coils is very large compared to the holding coil's magnetic circuit air gap. The ratio is about 3/8 inch for the closing gap and a few thousandths of an inch for the holding gap. Consequently, the holding coil is able to keep the contactor open with a very low current in the coil. The time delay period is varied by adjusting the magnetic air gap in the holding coil circuit: A small gap gives the greatest inductance and the maximum time delay.

The circuit of Figure 7.13 includes inductive time-limit acceleration and operates as follows: Closing the motor switch energizes the shunt field and FL coil. The FL coil closes the FL contacts in the UV coil circuit. Closing the control switch energizes the UV coil through the normally closed master switch reset contacts (provided the master switch is in the center position), the E STOP button, OL contacts, the UV coil, the PS contacts, and FL contacts. The UV coil closes the UV contacts.

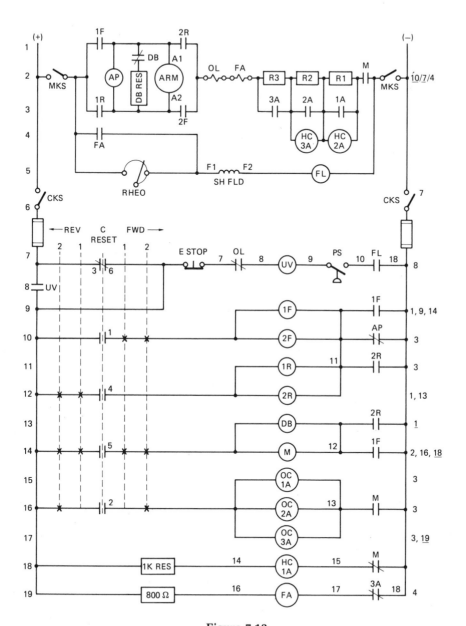

Figure 7.13

Closing the control switch also completes the circuit to HC1A and the FA coil. The FA tips close, shunting the field rheostat to provide for full field starting and maximum starting torque. Moving the master switch to the second point forward closes #1, #5, and #2 contacts in the controller and opens the MS RESET contacts. The #1 contacts complete the circuit through 1F and 2F coils through the normally closed AP contacts. The 1F and 2F contacts in the motor circuit close, setting the armature circuit up for forward operation. The 1F contacts in line 9 close to maintain 1F and 2F coils during operation. The 1F contacts in line 14 close, completing the circuit to M and DB coils and energizing them.

In the motor circuit, the DB contacts open and the M contacts close, completing the armature circuit and starting the motor through all the resistance. The voltage drop across R1 and R2 energizes HC2A and HC3A. The M contacts in line 16 close, completing the circuit to the 1A, 2A, and 3A coils. Although the operating coils are energized, these contactors cannot close because the holding coils have a voltage applied to them.

Normally closed M contacts in line 18 open to deenergize HC1A. After the flux decays, the 1A coil closes the 1A contacts in line 3, shunting R1 and HC2A. Now the flux produced by HC2A decays, permitting 2A to close. This shunts R2 and HC3A and after a time 3A closes, shunting R3 contacts to connect the armature across the line. Observe that a portion of resistance was removed from the armature circuit with the closing of the 1A, 2A, and 3A contacts.

Contacts in line 19 now open and deenergize the FA coil. After a time period, current decreases in the FA current coil and the FA contacts open to connect the rheostat in the shunt field circuit. Counter EMF decreases and causes an increase in armature current, reenergizing the FA current coil. The FA contacts close to shunt the rheostat and strengthen the shunt field. Because of the increase in field current, counter EMF increases, thus lowering the armature current. The FA current coil deenergizes and the FA contacts open. This happens repeatedly until the FA current coil fails to energize when the FA contacts open. At this point, armature speed should be very near the desired speed according to the rheostat setting.

To change the direction of rotation, the motor must come to a very low speed because of the AP relay.

The circuit of Figure 7.14 includes several of the control devices previously discussed and is presented here for the purpose of review. The circuit functions as follows: Closing the motor knife switch energizes the shunt field and the FL coil. The FL coil closes the Fl contacts in the UV coil circuit.

Closing the control knife switch energizes the UV coil, provided the master switch is in the center position and the remainder of the circuit is complete.

Moving the master switch to the fourth position forward closes the #1, #2, #5, #6, and #7 contacts in the master switch and opens the #3 reset contacts. The #2 contacts complete the circuit to the M and DB coils, energizing them. The M contacts in the power circuit close, and the DB contacts open. The M contacts in lines 15 and 18 close. The FA coil energizes and closes the FA contacts in line 4 to shunt the field rheostat to provide a full field for starting. The DB contacts in line 12 close and the 1F and 2F coils energize. The 1F and 2F main contacts in lines 1 and 3 close and the armature starts in the forward direction.

The armature circuit is as follows: From the positive line, through the MKS, 1F contacts, the armature A1 to A2, 2F contacts, OL coil, FA current coil, all the resistance M contacts, the MKS, to the negative line.

The 1F contacts in line 12 open to provide for electrical interlock. The 2F contacts in line 15 close to energize the P coil. The P coil closes the P main contacts to shunt the RP resistance. A high current flows through the 1AR coil and the 1AR contacts in the 1A coil circuit open to prevent the 1A coil from energizing. The P contacts in line 16 close. After a time delay 1AR contacts close to energize the 1A coil. The 1A contacts in line 3 close, shunting resistance R1. Again current flow increases owing to the change in resistance. Now the current surge is through the 2AR coil and the 2AR contacts in line 17 open. The

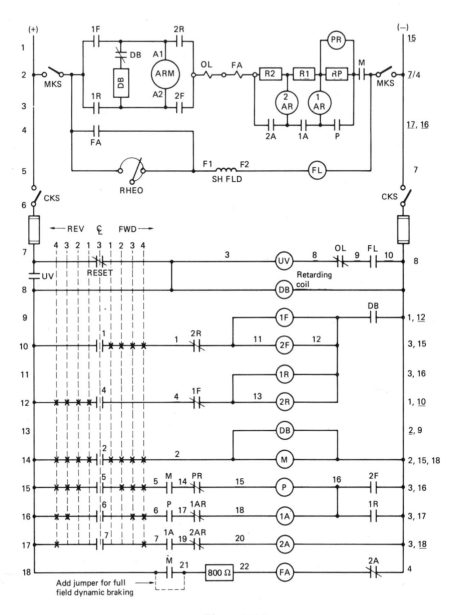

Figure 7.14

1A contacts in line 17 now close. Note that because of contactor design, the main contacts close before the auxiliary (control circuit) contacts close. During this brief interval, current is sent through the accelerating relay coil and the AR contacts open.

After a time delay, the 2AR contacts close to energize the 2A coil. The 2A contacts in line 3 close to shunt resistance R2. Again, current flow increases because of the change in resistance. Note that the FA current coil is in the armature current path. The field accelerating relay will function as described in the circuit of Figure 7.13.

If the motor is running across the line and is plugged, a high voltage will be applied to the armature circuit since the applied EMF and the counter EMF are of the same polarity. This causes a current flow higher than the normal starting current. The P main contacts open when the #5 controller contacts and 2F contacts in line 15 open. The current flow through RP causes sufficient voltage drop to energize the PR coil that is connected parallel to it. The PR contacts open. The P coil cannot energize until current through the armature circuit decreases to let the PR coil drop out. As soon as the armature stops turning forward (depending on relay adjustment), the PR coil drops out, the PR contacts close, and the acceleration begins as described above.

Chapter 8 Development of Control Circuits

ONLY THE MOST BASIC OF CONTROL CIRCUITS can be drawn as a complete system. Most circuits must be developed one step at a time, line by line, to meet the operating requirements of the system. Developing a control circuit is similar to writing a letter. The writer has the points in mind and proceeds word by word, line by line, until all the points are on paper. The same procedure applies to developing a control circuit. The developer should have each control function listed in the sequence of operation in order to provide for each function in the proper relationship to other functions as the circuit is developed.

There are two basic types of control circuits—two-wire circuits and three-wire circuits. These designations come from the number of wires each requires from an ordinary across-the-line motor starter to the control components. The three-wire control circuit uses momentary contact devices with a maintaining circuit provided by an auxiliary contact on the starter. Devices such as limit switches and float switches may be used to supplement the primary start–stop devices in various parts of the circuit, depending on the system requirements. The two-wire circuit uses a maintained contact device as the primary pilot control component. Any type of device that closes a set of contacts and holds them in the closed opposition for the required operating period may be used.

Figure 8.1 illustrates the basic two-wire and three-wire control circuits.

The purpose of this chapter is to show how these basic circuits can be expanded to achieve the necessary control of a motor or motors by adding push buttons or contacts of various control devices.

The schematic diagram is used for the development of a control circuit. To develop a control circuit, begin with two vertical lines to represent the power supply and add each

118

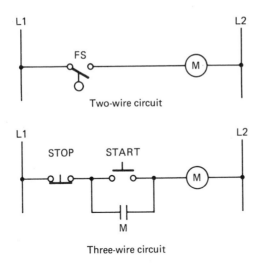

Figure 8.1

component between the lines to perform the desired function in the proper sequence.

As a general rule, devices that perform a stop function should be normally closed and are connected in series with the original stop button. Devices that perform a start function should be normally open and should be connected in parallel with the original start device. If several start devices are required to close before a circuit will operate, the devices should be connected in series with each other and in parallel with the original start device. Similarly, if more than one stop device must be activated in order to stop, the normally closed devices should be connected in parallel, then in series with the original stop device.

Each step of a required operating sequences should be added to the circuit individually. The circuit should then be checked to be sure a previous step was not interfered with when the latest step was added.

With good understanding of the fundamentals of control devices, you have the necessary information to learn to develop control circuits. To become proficient in developing control circuits requires much practice. The rewards of learning to develop control circuits are extremely valuable in analyzing and troubleshooting someone else's circuit, because in the development of circuits you will learn and understand the purpose of every device included in the circuit. This knowledge helps the troubleshooter to go immediately to a problem area when a circuit has failed.

To learn the step-by-step method of control circuit development, consider each circuit as a series of steps done at different times in order to achieve the overall desired result.

As an example, let us design the control for a pressurized coating oil system. An underground storage tank stores the oil which is to be used on a processing line. The oil is to be pumped through a filter from the storage tank into a pressurized tank. Here the oil is to be held at a controlled temperature and within certain pressure limits. It is necessary that the circuit include both manual and automatic modes of operation. Do not permit the pump to be started if the storage tank is empty. Provide for automatic air injection into the pressure tank when the tank pressure decreases. Obviously, we cannot immediately draw the circuit. The following steps are taken and a circuit is developed for each step.

Step 1. Draw a three-wire circuit to start the pump to fill the pressure tank in the manual mode. Provide overload protection. An operator will be required to observe the need to fill the tank, press the START button to fill the tank, and press the STOP button when it is full. See Figure 8.2.

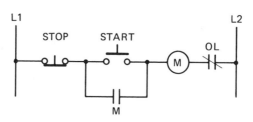

Figure 8.2

Step 2. Add the automatic mode of operation. This will require an automatic–manual, two-position selector switch and a normally open float switch. Use a maintaining circuit along with the float switch to permit the pressure tank to fill completely. See Figure 8.3.

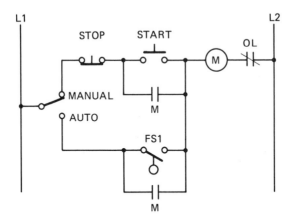

Figure 8.3

By examining the circuit we have just developed, we find that when operating in the automatic mode the pump will not stop when the pressure tank is full. A normally closed float switch mounted in the pressure tank and connected in the coil circuit will perform the STOP function. The circuit should be revised as shown in Figure 8.4.

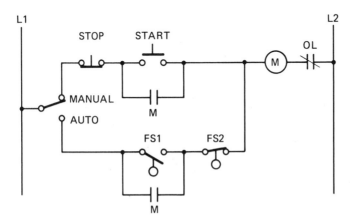

Figure 8.4

Step 3. Prevent the pump from starting if the storage tank is empty. To perform this function we can mount a normally open float switch in the storage tank and connect it in series with the coil. The float switch will close when a predetermined oil level is in the tank and open when the oil falls below this level. See Figure 8.5. FS3 is a normally open device that is included in the circuit as a permissive device. It is normally open and held closed during operation by an adequate oil level in the storage tank.

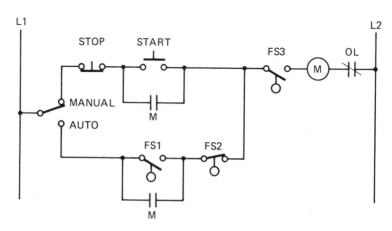

Figure 8.5

Step 4. Add the air pressure control devices. We should be aware that pressure in the tank will lower somewhat when the oil in the tank is in the lower limit. Therefore, air should be added to the tank only when the oil is in the upper limit and the air pressure is low. An electrically operated air valve (solenoid) is used. A set of normally open contacts on FS2, which operates on high level in the pressure tank, can be used in the air control circuit to serve as the START device. A normally closed pressure switch will serve as the STOP device. See Figure 8.6.

Step 5. Arrange temperature control switches to keep the oil in the pressure tank between 100 and 115°F. In the circuit shown in Figure 8.7, contactor CR1 will control an oil immersion heater. CR1 is controlled by two thermostats set for the high and low limits. If oil in the pressure tank drops below 100°F, TH1

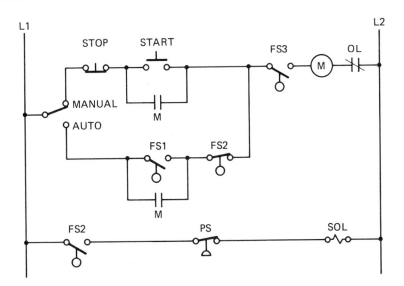

Figure 8.6

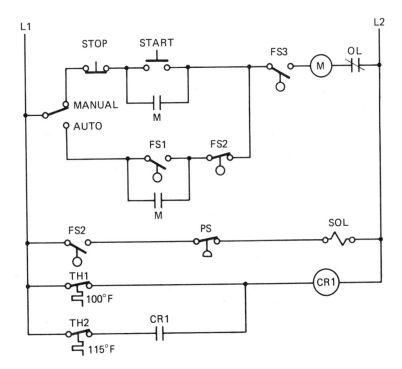

Figure 8.7

will close and CR1 will energize. The closing of CR1 main contacts energizes the heating elements (not shown). When the temperature exceeds 100°F, TH1 contacts open. CR1 is maintained through TH2 and CR1 contacts. TH2, set for 115°F, is closed and connected in parallel with the 100°F switch. When the temperature of the oil reaches 115°F, TH2 opens to deenergize CR1 and turn off the heater. When the oil temperature decreases, TH2 closes but CR1 does not energize until the temperature falls below 100°F to close TH1. See Figure 8.7.

The physical arrangement of the components is shown in Figure 8.8.

As another example, we design a control circuit to alternately energize two circuits, A and B. A motor-operated limit switch operates every 30 seconds and serves as the initiating device. This contact closes and remains closed for 30 seconds. It then opens and remains open for 30 seconds. For the first closing of the contact, circuit A should energize and remain energized until the switch opens. When the switch closes the second time, circuit B should energize. When the switch opens again, circuit B should deenergize. This cycle repeats itself.

It would be extremely difficult to make the final drawing for this circuit without going step by step through the development process. First let us list every action that is to take place in the proper sequence.

1. When the limit switch closes, circuit A is energized.
2. When the limit switch opens, circuit A is deenergized.
3. When the limit switch closes for the second time, circuit B is energized.
4. When the limit switch opens for the second time, circuit B is deenergized.
5. Return to step 1.

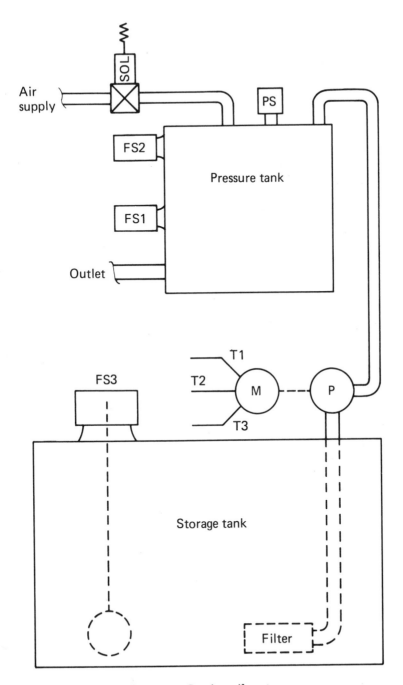

Figure 8.8 Coating oil system.

To develop the final circuit we proceed step by step.

Step 1. Represent circuits A and B and the initiating device. See Figure 8.9.

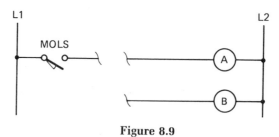

Figure 8.9

Step 2. Arrange to energize circuit A or circuit B, depending on the state of a control relay. See Figure 8.10.

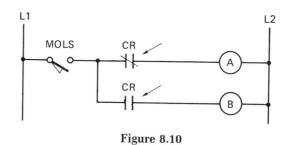

Figure 8.10

Step 3. Energize and maintain the CR relay. See Figure 8.11.

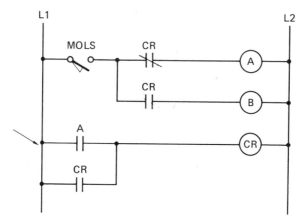

Figure 8.11

Step 4. Arrange to prevent B from energizing. See Figure 8.12.

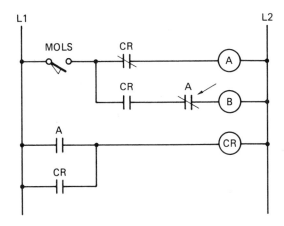

Figure 8.12

Step 5. Arrange to hold in A circuit. See Figure 8.13.

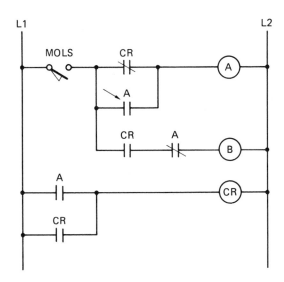

Figure 8.13

Step 6. Arrange for the next step in the sequence of control by deenergizing the CR coil when the B coil energizes. See Figure 8.14.

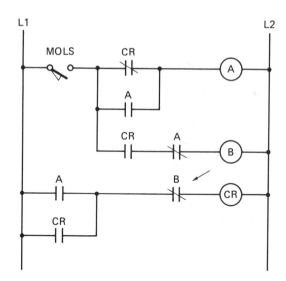

Figure 8.14

Step 7. Interlock the A coil circuit to prevent energizing it
when the CR normally closed contacts close.

Step 8. Provide a maintaining circuit for the B coil. See Figure
8.15.

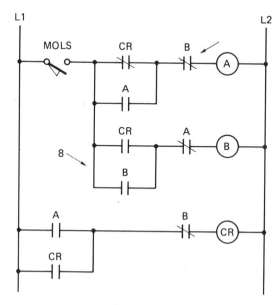

Figure 8.15

Development Exercises

1. Develop the control circuit to control a hydraulic cylinder to extend the ram from its retracted position, P1, to a predetermined position, P2. When the ram reaches P2 it is to stop and return to P1. Provide for returning the ram to P1 at any time in its forward movement.

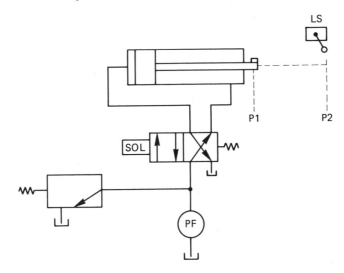

2. Develop the control circuit to control two hydraulic cylinders as follows. The cylinders are mounted horizontally and extend toward each other. Cylinder A extends from left to right and cylinder B extends from right to left.

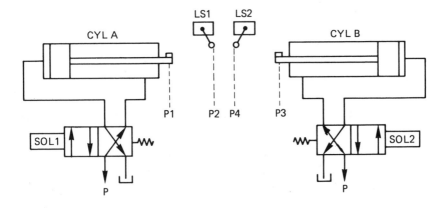

The #1 ram should extend from the retracted position, P1, to a predetermined position, P2. When it reaches P2 it should stop and return to P1. When the #1 ram begins to reverse, the #2 ram should move from the retracted position, P3. When the #2 ram reaches a predetermined position, P4, it is to stop and return to P3. Provide for returning either ram to its initial position at any time.

3. Design a control circuit for the hydraulic circuit shown below. Cylinder B is mounted to move horizontally from P3 to P4 and stop. Then cylinder A is mounted to move horizontally from P1 to P2 and stop. Cylinder B exerts increasing pressure against the workpiece until the desired pressure is reached. Both cylinders then retract together.

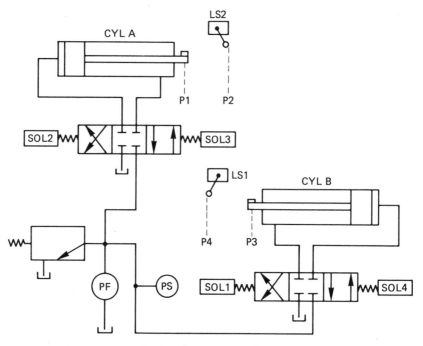

4. Design a control circuit to control two motors as follows. Pressing the START button starts motor #1. It continues to run until either the STOP button is pressed or an overload trips. When motor #1 stops, motor #2 should start. After running for 2 minutes, motor #2 should stop.

5. Design a control circuit to start two motors from a single

start–stop station simultaneously as follows. Pressing the STOP button or tripping an overload on motor #1 should stop that motor only. One minute later, motor #2 should stop. If an overload on motor #2 trips, both motors should stop immediately.

6. Design a control circuit to start three motors from one normally open push button. There should be a 15 second delay between starts as the motors start in sequence. The tripping of an overload on any motor starter should stop only that motor. If an overload should occur and the remaining two motors are stopped by pressing the STOP button, the original starting sequence must occur.

7. Design a control circuit for three conveyors arranged in tandem to convey material from a warehouse to a shipping dock. Conveyor #1 dumps onto conveyor #2, which dumps onto conveyor #3. Conveyor #3 empties into a container for shipping. The circuit should operate as follows:

   (a) One normally open push button should start the conveyors in sequence from last to first so that each conveyor is running before receiving additional material.
   (b) An overload on any of the motors will shut down the system.
   (c) A single STOP button should stop the conveyors in sequence from first to last.
   (d) Provide a 2 minute delay between the stopping of each conveyor so that it will be empty when it stops.

8. Design a control circuit to control two motors as follows:

   (a) Pressing a START button should start motor #1 and 1 minute later motor #2 should start.
   (b) After motor #2 has been running for 2 minutes, it should stop while motor #1 continues to run.
   (c) Pressing the STOP button should stop both motors.
   (d) An overload on motor #2 should stop both motors.
   (e) An overload on motor #1 should not affect motor #2.

9. Design the circuit for a control system to signal an alarm 15 seconds after the entrance door to a supply is opened. Permit the operator to set the alarm system inside the room and to leave the room without sounding the alarm. Also provide for the operator to reenter the room without an alarm.
10. Design the control circuit to start four motors in sequence from a single push button. If any of the overloads trip, the entire system should stop. Design the circuit to limit the current in the STOP button, the START button, each of the overload contacts, and the auxiliary contacts to the current drawn by one coil.

Answers to Development Exercises

1.

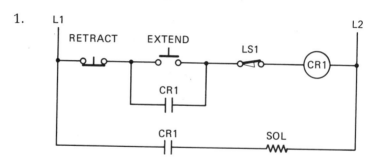

2.

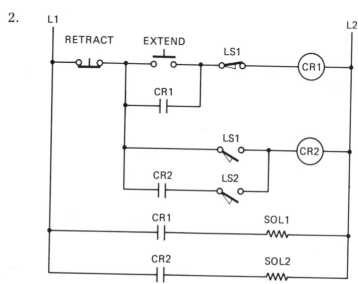

3.

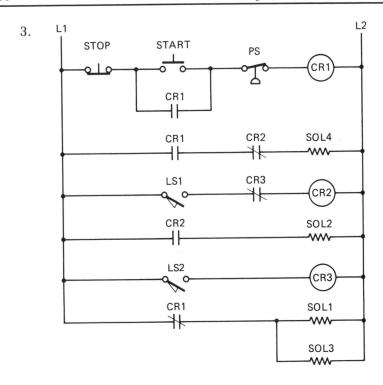

4.

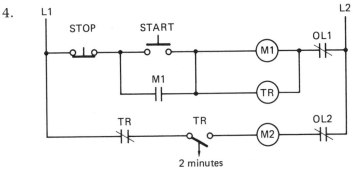

5.

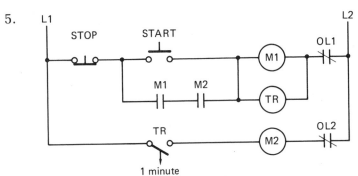

6.

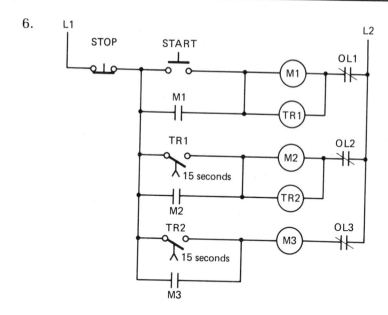

7.

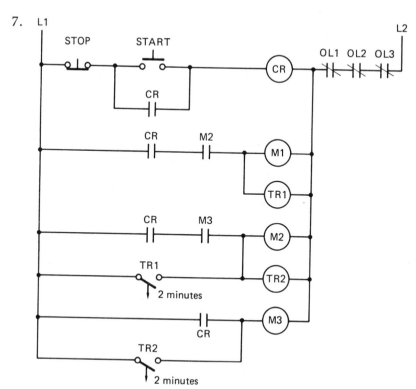

8.

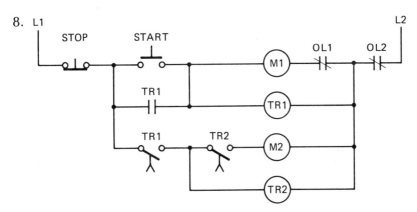

9.

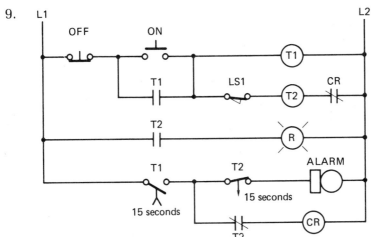

10.

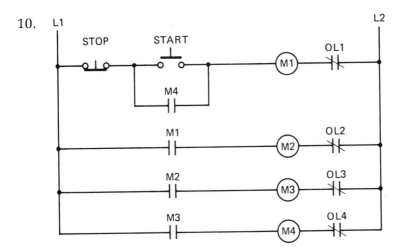

Summary

The examples and types of circuit discussed in this section do not begin to deal with the possible circuit variations that may be developed. However, they do explain the basic principles involved. Above all else, do not try to develop the complete circuit at one time. The main thing to remember in developing a circuit of your own is to proceed step by step to design each function until the entire circuit has been developed. Also, remember to check the entire circuit from the beginning each time a change or addition has been made. No matter how minor the change may seen, it could affect some function already included in your circuit.

While all control circuits consist of coils and contacts arranged to provide a programmed sequence, the operation of the contacts of various types of control device vary from one manufacturer to another. The basic principles do not change but various techniques are applied. It is beneficial and highly recommended for a new student in circuit development to review manufacturers' literature of the various control devices.

In addition to learning more about control components, practice in developing or modifying control circuits is also highly recommended for those who wish to become proficient in circuit development and troubleshooting. While the principles involved may seem simple, the only way to develop skills in applying the principles is through practice. Your skills will depend on how much you practice.

Chapter 9 Maintenance and Troubleshooting

TROUBLESHOOTING IS A FIELD of repair work that usually tells how well the student has learned the lessons. The principles involved in control functions, components, and circuit analysis, along with the basic laws of electricity, are all applied. The preceding chapters of this book were intended to provide the electrical troubleshooter with a broad knowledge base on which to build. Students often say that they need to know more about reading schematic diagrams and more about the use of a multimeter. These statements reveal a lack of understanding of the basic principles involved.

It is a simple matter to read a schematic diagram if the basic laws of electricity and the principles of control language are known. These factors have been presented throughout this book. While circuit diagrams can become involved and operating systems can be complex, the principles are neither. Concerning the multimeter, it can only confirm or disprove what you already suspect about a circuit. It is not capable of analyzing a circuit. A multimeter can only display the magnitude of the electrical information between the two points of the circuit to which the meter is connected. The troubleshooter must observe the reading and determine its meaning.

Your best tool when troubleshooting is your ability to think. Don't jump to conclusions. Have confidence in your ability. Learn how the equipment in your area is supposed to operate, both electrically and mechanically. Know before a failure occurs where pressure switches, limit switches, and other control devices are located.

The following material is intended to introduce you to troubleshooting techniques and is not presented as a comprehensive dissertation on troubleshooting. Use it to develop and enhance your own skills and techniques.

137

Finally, be careful. Observe all plant rules and regulations. Electricity can be dangerous. In addition to the hazards of electrical shock and electrocution, burns from an electrical flash can be devastating. Be careful when opening a circuit. The inductive kick that can occur when a circuit opens produces a voltage that is many times the voltage applied to the system.

Troubleshooting Hints

No matter how complete or expensive an electrical control system is, the components of the system begin to deteriorate as soon as they are installed and failure of some component in the system will ultimately result.

There are certain definite and logical methods or procedures in locating the source of trouble on electrical equipment. Experience indicates that in most cases where the exact trouble spot is not determined, it is because the troubleshooter has not applied his or her knowledge properly; that is, definite logical steps were not followed.

Blown fuses, overload contacts, open contacts, short circuits, burned out coils, and grounds are responsible for most electrical circuit failures. These problems should be relatively easy to find and correct.

Many of the more "sophisticated" systems fail because of some minor adjustment problem that requires more information than has been furnished to all the repair people. Records indicate that this type of failure is infrequent. The larger and more complicated system usually fails for the same reasons as the smaller and less complicated system: dirty contacts, open circuits, blown fuses, burned out coils, faulty grounds, broken limit arms, or some other mechanical aspect relating to the electrical operation. It is important to remember that these simple things happen more frequently; therefore, refrain from going immediately into complex adjustment procedures without first checking out the circuit for overloads, blown fuses, open contacts, and so on.

Troubleshooting can be generalized in three steps:

1. Determine the symptoms; that is, find out how it acts. When equipment is operating properly, you should find out how it is supposed to function.
2. Decide by logical reasoning what might be wrong. Try to isolate the problem to a section of the control.
3. Determine what has to be done to correct the problem.

The effective application of these rules requires an analytical mind trained in all aspects of control functions, components, circuits, and circuit analysis.

If we are troubleshooting an existing circuit, one that has been in service and operated properly, we can eliminate the possibility of faulty installation or design. Do not waste time by checking the drawing for design accuracy.

The first step—determine the symptoms—can best be accomplished by working with the machine operator and following the machine through its sequence to the point of failure. This should isolate the trouble spot to a relatively small section of the drawing. A thorough check of this section of the drawing should lead you to the problem. Keep in mind that the malfunction of some control component has caused the failure: However, grounds can cause the component failure so be sure to check the resistance to ground. Once the trouble circuit is located, check out each component in that circuit.

Remember that no matter how complex, control circuits are made up of only two things. Contacts that open and close a circuit, and coils that operate the contacts. Keep in mind that limit arms or other mechanical operating mechanisms can often fail mechanically and cause contacts to fail to operate. If the contacts open and close properly, the correct voltage will be applied to the operating coils to bring in further sequencing of the control circuit. If the control contacts do not operate properly, the proper voltage will not be applied to the coils. It should be relatively simple to determine where the voltage in a circuit is and to determine where the operation stopped sequencing.

Probably the single most important rule in troubleshooting is to remember to change only one thing at a time. No matter how poor the overall condition of the control is, it is usually the failure of one component that stops the operation. Sometimes more harm than good results from excessive parts changing and random adjusting.

If someone takes charge and directs the troubleshooting, much confusion can be eliminated. "Taking charge" means not only directing the work of other members of the maintenance crew but also directing the operator to perform certain operations so that the control or the machine can be observed. Care must be taken at all times to avoid injury to personnel or machinery. In some cases it is not practical to run the equipment when a part of the system has failed. The main function of the person in charge is to coordinate the efforts of the entire work group: Without this control, confusion usually results.

Remember, the operator knows the machine operation and can be an asset to you in your troubleshooting. Question the operator but don't challenge his or her operating ability. Many times the operator can direct you immediately to the trouble spot. I don't suggest that you change motors, contactors, and so on because the operator says they are bad (perhaps the motor isn't running because a wire is broken, not because the motor is bad). However, it would be wise to listen to the operator and check out any theories being offerd. Remember, the operator works the machine day after day and probably knows more about its unique operating characteristics than you do.

An electrical failure can result from a mechanical malfunction of control components as was previously mentioned. Limit switch trip arms are sometimes knocked out of proper adjustment or even broken off. Relays and contactors can become mechanically bound to prevent operation. Bearings can lock up. Mounting bolts can loosen or break. The possibilities of mechanical malfunction are unlimited so be sure to check for mechanical failure as well as electrical breakdown. Usually, the same troubleshooting procedures can be used to locate the problem. For instance, if you find the electrical circuit fails because a limit switch does not close, the next step should lead you to the limit switch where you may find a broken limit trip arm, loose base bolts, and so on. Also, if you find a burned out motor, be sure the machinery it drives is not defective. Many motors burn up because of bad bearings or gears in drive cases. Repeated tripping of the overloads can indicate a mechanical problem.

Anyone attempting to troubleshoot without a drawing and a meter is wasting time. Even if you know the general control sequence, it is unlikely that you know the order of the devices in the circuit, wire numbers, and other information necessary to be thorough in your troubleshooting. Furthermore, even though you know electricians who do not use a drawing and a meter in their troubleshooting, it is an indisputable fact that they could be even better at their job if these tools were added to their knowledge and ability in troubleshooting.

The following section is intended not as a complete set of instructions for troubleshooting but merely as a foundation on which to build good troubleshooting techniques. Each person must develop his or her own method of applying certain fundamentals. A thorough understanding of electrical theory is certainly an asset to the troubleshooter. (Perhaps a review of Ohm's law, Kirchhoff's law, Watt's law, and simple circuit analysis would improve troubleshooting ability.)

A portion of a motor control circuit is shown in Figure 9.1. The operator complained that the motor would run but would not accelerate. Upon checking the control panel, we found that

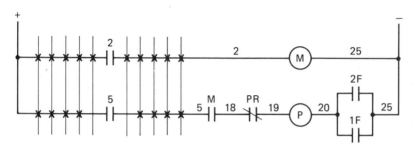

Figure 9.1

the M coil would energize but the P coil would not. Our task is to find out why the P coil will not pick up and to make the necessary repairs or adjustments.

Let me remind you to *take time to think*. Although one of the first things to do when troubleshooting a circuit is to check for blown fuses, overloads, circuit breakers, and so on, that would be an obvious waste of time in this situation since this type of failure would prevent the entire circuit from operating. Only when there is total failure is it necessary to begin at the power source. In other cases, find the last stop of the sequence that operated and begin there. In this example, we have found that since the M coil energized but the P coil did not the trouble must be in the P coil circuit. Any of the components or wires in that circuit could be defective.

Instead of randomly checking from one place to another, be systematic. Place a voltmeter across the P coil to find out if it has a voltage applied to it. If we read line voltage across the P coil but it will not operate the contactor, one of two things is probable: The coil is open or the contactor is mechanically bound. If we do not read line voltage across the coil, the circuit is probably open. Let us assume that this is the problem and see how we can find the open circuit. Connect the positive lead of the meter to the positive side of the line (a convenient place to connect is at the control disconnect switch) and the negative lead to wire 5 (probably the most convenient place to connect to wire 5 is the M tips). If line voltage appears here, we can assume that the MS 5 tips are open and that the remainder of the circuit is good. If we do not read a voltage across the MS 5 tips, the tips are probably making contact. It is possible that the tips are open and that there is another break in the line, but this would be unusual. Usually a circuit failure occurs when the first open occurs.

For the next step in checking out the P coil circuit, move the negative lead of the meter to wire 18 on the M tips. Again we have the same possibilities of readings. If we read line voltage, the M tips are probably open; if we do not read line voltage, the M tips are probably good. It may bother you that we say "probably" so many times. Can we ever be sure where the defective component is? Indeed we can. Suppose we get a voltage reading when we connect the negative meter lead to wire 18. This can only mean that the plus lead of the meter is on a plus voltage and the negative lead of the meter is on a negative voltage. To verify that the only break in the circuit is at the M tips, place the meter leads across the M tips. If we read line voltage there, we know that this is the only open in the circuit.

Now that we have found the problem, it is easily corrected. Clean the tips with a tip cleaner or by sanding with a fine grade of sandpaper. Control circuit contacts are usually silver plated and care should be taken that an excessive amount of silver is not sanded away.

If we had not read line voltage at the M tips, we would have proceeded across the circuit until we did get a voltage reading, and at that point we suspect trouble. We determine that it is the only open in the circuit by reading across it. Let me remind you that this is only one of many ways to look for defective circuits. Other techniques may be equally effective for you, but the important thing is to develop a systematic troubleshooting procedure.

Short Circuits

Suppose for some reason the P coil becomes "shorted." That is, the insulation on the coil windings breaks down and most of the coil's windings touch adjacent windings, causing the ohm resistance and the number of turns to approach zero. If the shorting out of the turns is excessive, current also becomes excessive and the control fuse blows. In the case of the P coil, the short is easy to locate. When sequencing the control, it is when the P coil energizes that the fuse will blow. An ohmeter can now be used to verify that the P coil is shorted. In other cases, where several coils energize simultaneously, it is more difficult to locate the problem. Coil leads can be removed and resistance checked until the shorted coil is located. Many times the bad coil can be spotted by its physical appearance—discolored from overheating with a distinct burnt odor. Don't fail to check for these obvious signs but remember that a coil will not always show visible evidence of failure.

In summary, when looking for short circuits:

1. Try to find where in the sequence the trouble occurs. That step of the control is the defective one.
2. Isolate components and check for resistance. Compare the coil resistance to similar coils you know to be good. To isolate control circuits it is sometimes necessary to remove control leads from several components. Be sure to mark leads so that they can be returned to their proper place. Sometimes this type of troubleshooting can overlap into the next turn and even though you may know where an unmarked wire goes, the next person on the job may not. Don't cause more problems than you already have. Again, take time to think.

Ground Faults

A ground fault is an unintentional connection to ground and should not be confused with intentional safety grounds made to motors, switch boxes, conduit, and so on. Ground faults occurring in an ungrounded distribution system present a very serious situation and should be cleared as soon as they are known to exist.

The most common type of ground detector in use is a simple light bulb arrangement as shown in Figure 9.2. When the system

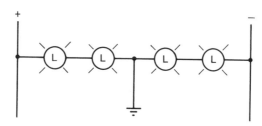

Figure 9.2

is ground free, all four lamps glow dimly and evenly since a small but equal voltage is dropped across each lamp. It should be apparent that if a ground fault develops anywhere in the system, the voltage drop across the lamps changes. For example, suppose we get a "dead" ground (no resistance to ground) on the negative line as shown in Figure 9.3. One pair of lamps will go out completely while the other pair will glow much brighter than before. Of course, if the connection to ground has resistance, the change in voltage across the lamps is not as dramatic and the change in illumination is also less pronounced. Once the lamps indicate a ground, a voltmeter can be used to determine the exact degree of the ground fault if you are interested.

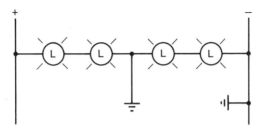

Figure 9.3

Ground detector lamps can do nothing to prevent or clear a ground fault. They only serve to indicate that the system has a ground. Ground detector relays are sometimes used to shut down a system when it develops a ground. The contacts of the relays are placed in the control circuit to shut down all parts, or any one part, in case of a ground. An example of this circuit is shown

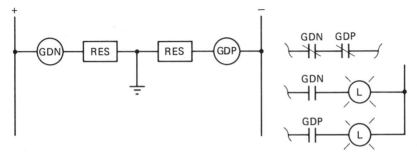

Figure 9.4

in Figure 9.4. The relays can be used to turn on indicating lamps or other aids in troubleshooting.

Grounds, unless accompanied by a short or open circuit, are generally located by the process of elimination. The ground circuit may be located by pulling switches to isolate equipment. Once the ground lights indicate a no-ground condition, the general area of the ground has been determined. Care must be taken when pulling switches to see that no switch is pulled under load.

Multiple grounds may be located by opening switches one at a time and leaving them open until the ground detector indicates normal. Then, with the grounded switch left open, the remaining switches should be closed one after the other until another ground is indicated. The switches of the grounded circuits should be left open and the procedure continued until every switch is tested. It would be to your advantage to obtain a listing of the equipment operating on various distribution systems in your area.

Once the ground is determined to be on a certain machine, the exact location of the ground can be determined by isolating various components on that machine and reading resistance to ground until the grounded component is determined.

A more sophisticated method of locating grounds, which is also more efficient, is the *ground fault locator*. This instrument sends a traceable signal through the system and allows the operator to trace the signal to ground. One of the biggest advantages of this instrument is that it prevents the necessity of shutting down equipment while looking for the ground fault.

In conclusion, troubleshooting electrical equipment often requires all of the commonsense knowledge of electricity you have acquired through your academic studies, but more often, the trouble is with fuses, overloads, dirty tips, and so on. Let me again emphasize the importance of the phrase *take time to think*. If you overlook a defective component early in your work you may begin on a needless, frustrating search for what should be a simple solution.

Contact Care

Any loose electrical connection will eventually cause trouble. A loose or open circuit can result in lost production time because it is sometimes difficult to find. A loose connection can cause a poor contact of high resistance. High resistance produces heating, and the increased temperature causes oxidation. The effect is always cumulative: Heating increases until the parts overheat, deteriorate, or cause an electrical fire. Loose connections, on thermally operated devices, may cause a relay to trip and stop a motor when it is not overloaded.

The bolts that hold contacts in place should always be tight. Normal expansion and contraction of metals resulting from temperature changes or excessive vibration may cause bolts to loosen.

Few contactors close without some bounce or rebound. This is because the reaction of the contact springs as they are compressed provides the final contact pressure. When the contacts bounce, they separate. At this time the contacts are carrying current and when they separate, an arc is created. The arc may cause sharp projections of burned or roughened contact surfaces to weld together. The contacts will not open when next required to do so. Other causes of contact welding are excessive currents, insufficient contact pressure, sluggish operation when either closing or opening, and momentary closing of contacts.

Every time contacts open or close they are subject to mechanical wear and electrical erosion. The reason for this is that most contacts close with a rolling movement combined with a wiping action. Although this ensures good contact and confines the arcing, the result is erosion to the tips of the contacts. Both conditions cause wearing of the contact materials. Contacts, therefore, are items that may require considerable maintenance, depending on the operating conditions. The actual mechanical wear of contacts that operate frequently may be more serious than the electrical erosion caused by arcing.

Because of wear, both mechanical and electrical, the pressure decreases. This affects the current-carrying ability of the contacts and will cause overheating of the contacts. A small contact with suitable pressure will carry current with less heating than a large contact with little pressure. Provisions are made for the wearing of the contacts when the original designs are made but replacements will eventually become necessary. Manufacturers furnish information on correct contact pressures for their devices. Contact pressures should be maintained within prescribed limits.

With the contactor coil energized and the contact sealed closed, observe the clearance of the moving contact from its stop. It should not be less than 1/64 inch. Renew contacts before this limit is reached. Replacement should always be made in pairs. Because of the wearing of contact surfaces, the probability of a mixture of old and new parts operating badly is very high. The extra time and expense spent in replacing both contacts will pay many times over in operating life.

Copper Contacts

Copper as a contact material has been used for many years, and when properly designed into a device, it works satisfactorily. Copper-contact surfaces are subject to oxidation and high resistance and, in most devices, must have proper wiping and cleaning action to maintain normal operating temperatures.

All contacts should be kept clean, especially copper contacts. The discoloration that appears on copper is not a good electrical conductor. It increases the contact resistance and often is the cause of serious heating. When contacts are renewed, it is important to clean the new contact and the surface against which it is mounted. The slight rubbing action and erosion that occurs during normal operation will generally keep the contact surfaces clean enough for service. Copper contacts that seldom operate will readily accumulate the thin discolored film that can cause heating.

Silver Contacts

Fine silver contacts, when properly applied, require lower contact force and are less subject to high contact resistance. They require substantially no cleaning or servicing, since the normal oxides are easily broken down and are of low resistance. Substantially, no wipe or roll is required in the low to medium current ranges.

The dense discoloration that soon appears on clean silver is actually not a good conductor, but the surface film breaks down easily. Experience has proved that it is not necessary to keep silver contacts clean.

Sintered or Composition Contacts

In higher-current applications, silver compositions and/or sintered contacts are used to obtain high conduction, low temperature, low erosion, and less tendency to stick. Wiping action is not required. Under normal conditions, maintenance is not required.

Miscellaneous Contacts

Slightly roughened surfaces that appear during normal operation will provide better contact area than smooth surfaces. Contacts with surfaces comparable with very coarse sandpaper may be in good condition. Contacts in use may become a purplish dark gray to brown, which is normal. Usually only the periphery of the contact need be cleaned. Contacts, for the most part, are self-cleaning, depending on the design and application of the material. When the contact force is on the low limit and the contact remains closed for long periods (which does not allow the self-cleaning action), the contacts are more subject to overheating, and a reddish-brown hard copper oxide forms which is of high resistance. This type of surface should be removed. If and when such conditions arise, a fine file can be used, followed by medium sandpaper— *NOT EMERY CLOTH*. Do not change shape by filing or grinding.

Do not oil contactor or relay bearings unless lubrication is a manufacturer's requirement.

Since the correct operation of the contactor depends primarily on its being completely clean and free from foreign material, no oil or lubricant of any kind should be used on the bearings unless the manufacturer's literature specifies lubrication. Generally, contactor bearings are designed to require no lubrication. If lubricated, the accumulation of oil and dirt may cause sluggish mechanical action that will impair the arc rupturing qualities of the device or cause welding of the contacts.

When contactors are required to open circuits carrying currents that are difficult to interrupt, they are designed with arc rupturing features. The arc rupturing parts must be in a definite position with respect to the contacts. The interrupter should always be returned to the proper position if removed for any reason.

The fine strands of the flexible shunts that carry current around the contactor bearings sometimes fray or break where the shunt bends. If very many strands break, the unbroken ones will eventually overheat and fail. Frayed shunts should be replaced promptly with new ones.

Coils provide the electromagnetic force that causes the contacts of relays and contactors to operate. Series coils generally carry heavy currents and have few turns of heavy copper. Shunt coils have many turns of fine wire.

Shunt coils for ac devices must close them at 85% of the rated voltage. Coils for dc devices should close them at 80% of the normal voltage. Any coil should withstand 110% rated voltage without damage. Measurement of operating voltages should be made at the coil terminals and not at the source of the supply voltages. A coil with an open circuit obviously will not operate. If some of the turns of a dc coil become short circuited, the resistance of the coil will be reduced and more current will flow. The increased current causes higher operating temperatures and frequently results in coil burnout. Short-circuited turns in an ac coil result in a transformer action that permits a high current flow at the point of the short circuit. Immediate burnout usually follows the short circuiting of an ac coil, whereas a short-circuited dc coil may continue to operate for an indefinite time, until total insulation failure results.

Coils should be operated at the rated voltage. Overvoltage shortens coil life because it operates the contactor or relay with excessive force and causes more mechanical wear and bounce. Undervoltage on coils causes contactors and relays to operate sluggishly. The contact tips may touch, but the coil may be unable to close the contacts against the contact spring pressure.

The impedance of a magnetic circuit having a large air gap is much lower than the impedance of the same magnetic circuit with no air gap. The current drawn by the coil of an ac magnet is therefore much greater before the contactor operates. Since the closing time is short, ac coils are designed to withstand continuous energizing under closed conditions. An ac coil will soon overheat if energized with a large air gap in the magnetic circuit. Contactors with ac coils should not be blocked open when testing. Direct current coils are not subject to these conditions because the coil currents do not vary with the air gap.

Because the voltage on the coil of an ac contactor or relay repeatedly passes through zero, the holding power follows. The voltage, however, is soon effective in the opposite direction, and the device is again pulled closed. This operation causes a humming noise in any ac-operated device and a decided chattering noise in a defective unit. This chattering is eliminated by the use of a shading coil which provides sufficient holding power to keep the device closed. Any dirt in the pole-face area introduces a greater air gap when the unit is closed and results in a noisy operation.

Direct current coils are not subject to a repeated zero-voltage condition. Therefore, dc-operated devices are always quiet. For this reason, ac current-carrying contactors equipped with dc-operating coils will operate quietly.

When operating coils are deenergized, residual magnetism remains in the magnetic circuit. It is sometimes strong enough to hold the device closed after the coil is deenergized. This condition occurs most frequently on small devices on which contact spring pressures and moving parts are light. It can be avoided by adding a nonmagnetic shim to the magnetic circuit. Should magnetic sticking occur, examine the device to see if the nonmagnetic shim is missing.

Commutator Maintenance

The commutator is a vital part of every dc machine. How well the machine performs depends largely on how well the commutator is maintained. A commutator cannot function as it should unless the brushes make good electrical contact. This requires a smooth, cylindrical surface that runs true with its center. If you will consider the speeds at which some machines operate, and that the brushes must ride on the commutator surface, you will understand why this is true. To maintain a commutator properly you must be able to determine when a faulty condition is starting and prevent it from worsening. To do this, you must be aware of various surface conditions and how they affect commutation. Figure 9.5 shows a cutaway section of a commutator.

A surface film of carbon, graphite, copper oxide, and water vapor is deposited on the commutator by the electrochemical action of the brushes and commutator as the machine operates. The brushes begin to establish the characteristic film as soon as current flows. The film is affected by temperature, atmosphere, and grade of brush. Humidity, chemical contamination, abrasive dust, oil vapors, and many organic materials greatly impact the commutator film. The machine manufacturer selects the proper brush grade depending on the application of the machine. Once

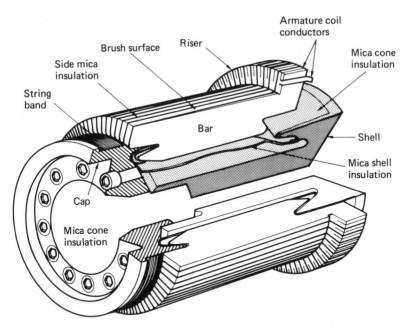

Figure 9.5 Cutaway section of a commutator.

the correct grade has been established, use only this grade. The commutator surface film is a critical factor in satisfactory commutation and in maintaining minimum brush and commutator wear. Changes in color from dark brown to copper are not unusual provided the surface is smooth and polished. A new shiny copper finish is not desirable. The commutator should look like an old penny and the surface should not be disturbed.

The development of a color pattern on the commutator bars is not a concern as long as the pattern is uniform around the entire commutator. Nothing needs to be done unless damage exists.

Possible causes of commutator damage are the following:

1. Excessive current load on the machine.
2. Electrical adjustment off.
3. Wrong brush grade.
4. Rough commutator surface.
5. Contaminated atmosphere.
6. Incorrect brush spacing.
7. Incorrect brush shift.

When the faulty condition is known, make the necessary correction and the commutator condition should remedy itself. Some commutator resurfacing may be required but should be done sparingly.

Arcing between the brush and commutator segments results in an opaque darkened surface or black deposit on the bars. If prolonged, this condition can cause surface etching. Etching usually occurs at the trailing edge of the commutator bar. If etching is not corrected, flat spots will develop on the commutator. Flat spots may also be caused by bad or worn bearings.

A commutator surface defect known as *threading* shows up as narrow, circumferential lines or grooves cut into the commutator surface. These lines vary from merely a discolored surface to a rough, grooved surface. Brushes will wear to fit the grooves. Contact between the brush and the commutator can be lost when the commutator shifts because of end play. Sparking and poor operation may result. Possible causes of threading are the following:

1. Operating the machine for extended periods with no current in the brush.
2. Abrasive foreign material embedded in the bottom of the brush. This could be copper or mica particles from the commutator.
3. The wrong grade of brush. When threading is pronounced, the commutator should be resurfaced. Mild cases of threading may be repaired with hand stoning. Severe cases should be ground or turned.

Overheating of the commutator can cause commutator bars to expand, with some of the bars rising above the others. Holding a motor at standstill with current applied can cause the bars under the brushes to overheat and expand. When rotation is resumed, the high bars can kick the brushes up so that they are not in contact with the commutator. Sparking will result. Severe deterioration may follow and result in flashover. If flashover occurs, clean the commutator, the brush holders, the bands, and other adjacent parts with a cloth and a solvent. Examine the faces of the brushes and replace damaged brushes.

If a commutator surface is merely smudged, you can clean it by polishing with canvas. If the commutator is rough, crocus cloth or fine sandpaper may be used. Do not use emery cloth. The abrasive particles scratch the surface and are conductive. Particles that lodge between commutator segments will lead to short circuits.

When working on a commutator, wear gloves and proper eye protection to safeguard against flying particles and possible flashover. A respirator should be worn during stoning to avoid inhaling abrasive dust.

Appendix A Electrical Troubleshooting Charts

Trouble	Possible Cause	Suggested Correction
ac Synchronous Motors		
Motor will not start	1. No supply voltage	Reset circuit breakers; replace open fuses
	2. One or more phases open	Replace fuse(s); check for defective wiring
	3. Mechanical overload	Correct cause of overload
	4. Insufficient voltage	Check relay contacts; check voltage supply and increase if possible
Motor runs hot	1. Mechanical overload	Check motor rating; reduce mechanical load
	2. Improper ventilation	Check for proper motor ventilation; clean all motor vents
	3. Shorted or open coils in motor	Repair/replace all defective motor coils
	4. High line voltage	Check and adjust supply voltage
	5. Stator grounded	Clear ground; replace stator
	6. Incorrect field current	Adjust current to proper level
	7. Rotor off-center	Check bearing wear; check concentricity of rotor
	8. Rotor in contact with stator	Check bearing wear; check concentricity of rotor
ac Synchronous Motors		
Incorrect speed	1. Improper frequency	Check to ensure correct frequency; low = slow speed; high = fast speed

Trouble	Possible Cause	Suggested Correction
Improper synchronization	1. Mechanical overload	Check motor rating; reduce mechanical load
	2. Field coils open	Repair/replace defective field coils
	3. No exciter voltage	Check voltage and adjust to correct level
	4. Improper field rheostat setting	Check rheostat and adjust to correct setting
Excessive motor vibration	1. Motor out of synchronization	Make necessary adjustments as listed above
	2. Open armature coil	Repair/replace coil
	3. Open incoming phase	Check incoming supply: fuses, wiring, etc.
	4. Misalignment of motor	Check alignment; realign if necessary

Single-Phase Motors

Trouble	Possible Cause	Suggested Correction
Motor will not start	1. No supply voltage	Check for open breaker or fuses and damaged wiring
	2. Defective controls	See control troubleshooting section
	3. Defective windings	Check for open or shorted windings; repair/replace defective components
	4. Defective capacitor	Replace open or shorted capacitor
	5. Mechanical overload	Check motor rating; reduce load as required
Motor runs hot	1. Mechanical overload	Check motor rating; reduce load as required
	2. Defective control circuit	See control troubleshooting section
	3. Incorrect line voltage	Adjust line voltage to correct level
	4. Incorrect line frequency	Adjust frequency to correct cycle rate

Trouble	Possible Cause	Suggested Correction
	5. Shorted stator coils	Repair/replace damaged coils
	6. Rotor contacting stator	Defective bearings; rotor is off-center in stator
	7. Worn bearings	Replace bearings
	8. Misalignment	Check and realign if necessary
	9. Improper ventilation	Clean vents; provide additional ventilation

Single-Phase Motors

Trouble	Possible Cause	Suggested Correction
Motor runs slow	1. Mechanical overload	Check motor rating; reduce as necessary
	2. Incorrect voltage	Adjust as necessary
	3. Incorrect frequency	Adjust as necessary
	4. Open rotor bars	Repair/replace as required
	5. Defective stator coils	Repair/replace as required

Three-Phase Squirrel Cage Motors

Trouble	Possible Cause	Suggested Correction
Motor will not start	1. No line voltage	Check for open circuit breakers or blown fuses
	2. One or more phases open	Check fuses, breakers, or damaged wiring
	3. Mechanical overload	Check motor rating; reduce load as required
	4. Rheostat open	Repair/replace rheostat
	5. Incorrect brush tension	Ensure sufficient tension to maintain good contact
	6. Rotor coils open	Repair/replace defective coils
Motor runs hot	1. Mechanical overload	Check motor rating; reduce load as required
	2. Defective stator coils	Repair/replace coils
	3. Incorrect line voltage	Adjust voltage as required
	4. Incorrect line frequency	Adjust frequency as required

Trouble	Possible Cause	Suggested Correction
	5. Line has one or more phases open	Check fuses, breakers, or damaged wiring
	6. Grounded stator coil(s)	Repair/replace coil(s)
	7. Misalignment of motor	Check and realign if necessary
	8. Uneven air gap around rotor	Rotor is eccentric; bearings worn
	9. Worn bearings	Replace bearings
	10. Inadequate ventilation	Clean motor vents; provide additional ventilation
Motor runs slow	1. Mechanical overload	Check motor rating; reduce load as required
	2. Improper line voltage	Adjust as necessary
	3. Improper line frequency	Adjust as necessary
	4. Damaged rotor coils	Repair/replace coils
	5. Damaged stator coils	Repair/replace coils
	6. One or more line phases open	Check for open fuses, breakers, or damaged wiring

dc Motors

Trouble	Possible Cause	Suggested Correction
Motor does not start	1. No supply voltage	Check circuit breakers; supply fuses; repair any damaged wiring
	2. Mechanical overload	Check motor rating; reduce load if necessary; replace worn bearings on armature
	3. Improper brush–armature contact	Adjust spring tension on brush holders
	4. Field coil open	Repair/replace field coil
	5. Armature circuit open	Repair/replace defective part of the circuit
Motor runs hot	1. Mechanical overload	Check motor rating; reduce load if necessary
	2. Defective armature coils	Repair/replace defective coils
	3. Brushes positioned off the neutral plane	Reposition brushes
	4. Incorrect line voltage	Adjust voltage to proper level

Trouble	Possible Cause	Suggested Correction
	5. Defective field coils	Repair/replace defective coils
	6. Improper ventilation	Clean motor vents; provide additional ventilation if necessary
Motor runs fast	1. Line voltage too high	Adjust supply voltage as required
	2. Improper field connections	Change connections as required
	3. Defective shunt field	Replace shorted or open coils
	4. Improper shunt field rheostat setting	Readjust as required
Motor runs slow	1. Low supply voltage	Adjust supply voltage as required
	2. Mechanical overload	Check motor rating; reduce load if necessary
	3. Incorrect current in armature circuit	Check for shorts or opens in the armature circuit
	4. Starting resistance not removed from circuit	Check the control circuit
	5. Brushes positioned off the neutral plane	Adjust the brushes as necessary
Sparking at the motor brushes	1. Contamination between brushes and commutator	Clean with solvent; polish and seat as required
	2. Wrong brush grade	Check for manufacturer's recommended grade
	3. Incorrect brush tension	Adjust brush tension as required
	4. Brushes positioned off the neutral plane	Adjust the brushes as required
	5. Open in armature circuit	Repair/replace defective component
	6. Eccentric armature	Grind if possible; otherwise replace armature
	7. High mica and/or commutator bars	Undercut mica and recut bars; otherwise replace armature
	8. External vibration	Eliminate vibration source; replace defective bearings; align and tighten motor

Trouble	Possible Cause	Suggested Correction
	9. Reversed connection on commutating coil or short in armature circuit	Properly connect commutating coil; repair/ replace defective components

Commutator Conditions

Trouble	Possible Cause	Suggested Correction
Marking, etching, flat spots	1. Excessive loads	Check motor rating; reduce loads as necessary
	2. Brushes positioned off the neutral plane	Adjust brushes to proper position
	3. Wrong grade of brushes	Check manufacturer's recommendations for correct brush grade
	4. Rough or uneven commutator surface	Stone or smooth commutator surface
	5. Contaminated environment	Filter all air flow into the motor
Threading of commutator bars	1. Low current density of brushes	Increase load or change brush grade
	2. Abrasive materials imbedded in bottom of brush	Change of brushes; remove source of abrasive material
	3. Wrong grade of brush	Check manufacturer's recommendations for proper brush grade
Copper dragging over mica	1. Environmental contamination	Filter air circulating through motor
	2. Copper imbedded in brush	Change brushes; remove source of copper
	3. Hard spot in brush	Change brushes
	4. Wrong grade of brush	Check manufacturer's recommended brush grade
	5. Excessive vibration	Remove source of vibration; check bearing wear; check motor base bolts for proper torque

Trouble	Possible Cause	Suggested Correction
Banding of commutator surface	1. Excessive brush film	Change brush grade
	2. One brush of set is wrong grade	Consult manufacturer's recommendation for correct brush grade
	3. Hard spot in brush	Replace brush
Commutator out of round	1. Bent armature shaft	Change armature shaft
	2. Worn or defective bearings	Change bearings
	3. Commutator machined incorrectly	Remachine commutator
	4. Wide temperature variations and high speed	Stabilize environmental temperatures
Breaks in commutator surface	1. Handling damage	Use proper handling procedures
	2. Commutator struck by foreign object	Repair damage; guard against further damage
	3. Brush holder in contact with armature	Repair damage; raise brush holder to proper position
High commutator bars	1. Loosened or hot bars	Prevent motor sitting still while energized; prevent handling damage or incorrect installation of bars
High mica	1. Normal wear of commutator bars	Undercut mica

Magnetic Controllers (ac)

Noisy magnet	1. Defective shading coil	Repair/replace shading coil
	2. Magnet faces worn	Repair/replace worn magnet parts
	3. Contamination on magnet	Clean magnet faces
	4. Low line voltage	Adjust line voltage to proper level

Trouble	Possible Cause	Suggested Correction
Broken shading coil	1. Weak spring pressure	Check spring for overheating; replace damaged springs
	2. Wrong coil	Replace with correct coil
	3. Overvoltage	Adjust voltage to correct level
Frequent operating coil failure	1. Overvoltage	Adjust voltage to correct level
	2. High environmental temperatures	Ventilate area surrounding controls
	3. Magnet not sealing in	Remove contamination; check spring tension; check voltage level
	4. Excessive jogging of relay	Check duty cycle of relay
	5. Noisy magnet	See above
	6. Metallic contamination on relay	Remove contamination and protect relay from further contamination
	7. Moisture on relay	Remove source of moisture
Excessive magnet wear	1. Overvoltage	Adjust for correct voltage
	2. Broken shading coil	Repair/replace shading coil
	3. Wrong coil	Change coil
	4. Incorrect spring tension	Check for overheated spring; install correct spring
	5. Noisy magnet	See Noisy magnet section

Magnetic Contactors ac and dc

Trouble	Possible Cause	Suggested Correction
Blowout coil overheats	1. Excessive current	Check for proper current level
	2. Coil too small	Install proper size coil
	3. Loose connections on coil and/or tip	Tighten all connections
	4. Tip overheating	Check for loose connections, weak spring tension, high current

Trouble	Possible Cause	Suggested Correction
	5. Excessive jogging of relay	Reduce operational frequency
Excessive wear of arc chutes	1. High-current interruptions	Check current ratings
	2. Excessive vibration	Remove source of vibration
	3. Excessive moisture	Remove source of moisture
	4. Improper handling and/or assembly	Ensure good handling and installation practices
Relay fails to pick up	1. Incorrect voltage	Check supply to ensure correct voltage
	2. Defective coil	Check for open or shorted coil
	3. Mechanical binding of relay	Check for free mechanical operation of relay
	4. Excessive magnet gap	Replace worn relay components to restore proper magnet gap
	5. Wrong coil	Check for correct coil application
	6. Open in coil control circuit	Check for proper voltage at coil

dc Motors

Trouble	Possible Cause	Suggested Correction
Broken flexible shunt	1. Normal wear	Replace as needed
	2. Improper installation	Use proper installation procedures for maximum shunt life
	3. Poor environmental conditions	Protect shunts from corrosive elements
	4. Burnt from excessive arcing	Check for proper contactor loads and good arc chutes
Contactor fails to drop out	1. Contact tips welded together	Change contact tips
	2. Mechanical binding of contactor	Repair/replace defective contactor components
	3. Contamination holding relay closed	Clean contactor and check for free operation

Trouble	Possible Cause	Suggested Correction
	4. Contactor held closed by arc chute	Check for correct alignment of contactor
	5. Control circuit energized	Check for fault in circuit
Insulation failure	1. Environmental contamination	Protect from moisture and corrosive fumes
	2. Overheating	Provide adequate ventilation; reduce cycle frequency
	3. Current and voltage surges	Check control circuit for proper operation
	4. Shorted coil	Replace coil
Coil sticks occasionally	1. Environmental contamination	Protect relay from all contamination
	2. Worn components in relay	Change all worn parts
	3. Mechanical bind of relay	Check for free operation; replace worn components
	4. Misalignment of relay	Realign and check for correct operation
Rapid mechanical wear	1. Environmental contaminates wearing relay components	Protect relay from environment; change worn parts
	2. Wrong relay for the application	Use continuous duty relays for constant cycling

Timing Relays

Trouble	Possible Cause	Suggested Correction
Relay sticking	1. Dirt in relay	Clean and replace worn parts; prevent dirt accumulations
	2. Improper adjustment	Check relay for correct settings
	3. Mechanical bind of relay	Check for free operation; replace worn components

Trouble	Possible Cause	Suggested Correction

Timing Relays

Trouble	Possible Cause	Suggested Correction
Excessive mechanical wear	1. Abrasive contamination	Remove contamination and protect relay from additional accumulations
	2. Wrong relay	Use continuous duty relays for heavy-duty applications

Overcurrent Relay

Trouble	Possible Cause	Suggested Correction
Low trip	1. Wrong coil installed on relay	Check coil application; install correct coil
	2. Relay assembled incorrectly	Reassemble following manufacturer's instructions
	3. Defective coil	Replace coil
High trip	1. Mechanical bind of relay	Repair/replace binding components
	2. Wrong coil installed on relay	Check coil applicataion; install correct coil
	3. Defective coil	Replace coil
	4. Relay assembled incorrectly	Reassemble following manufacturer's instructions
Fast trip	1. High temperature	Cool ambient temperature; install larger relay
	2. Wrong heaters	Check manufacturer's recommendations; install correct heaters
Slow trip	1. Mechanical bind	Repair/replace relay
	2. Contamination	Clean and protect relay
	3. Low ambient temperature	Change relay coil
	4. Wrong heaters	Check manufacturer's recommendations; install correct heaters

Trouble	Possible Cause	Suggested Correction
Solenoids		
Noisy operation	1. Low supply voltage	Adjust supply to proper setting
	2. Mechanical bind	Repair/replace source of mechanical bind; reduce the shifting force required
	3. Magnet faces misaligned	Align and properly install magnet components
	4. Contamination in relay	Remove source of contamination; protect relay from further contamination
	5. Defective shading coil	Repair/replace shading coil
Defective shading coil	1. Overvoltage	Reduce to proper level
	2. Mechanical load too small; heavy slamming	Check for correct spring; repair/replace worn parts
	3. Wrong coil	Check manufacturer's recommendations; install correct coil
	4. Low line frequency	Check and repair problems
Coil failure	1. High ambient temperature	Ventilate area; install larger coil
	2. Magnet not sealing in	Remove foreign material; replace worn parts
	3. Line voltage too high	Reduce line voltage
	4. Metallic contamination in relay	Remove contaminants; replace worn parts
	5. Wrong duty cycle	Use continuous duty relays for heavy-duty application
	6. Mechanical bind or overload	Remove obstruction; replace worn components; free relay motion
	7. Insufficient mechanical load (slamming of relay)	Check for broken components and overvoltage
Magnet face wear	1. Slamming of relay	Replace broken or missing spring; reduce overvoltage; replace defective relay linkage

Trouble	Possible Cause	Suggested Correction
	2. Broken shading coil	See Shading coil section
	3. Wrong coil	Check manufacturer's recommendation for coil
	4. Noisy relay	See Noisy relay section
	5. Magnet misalignment	Repair/replace worn components; correct alignment
Solenoid fails to pick up	1. Mechanical overload	Reduce force required to energize relay; repair/replace worn components
	2. Incorrect supply voltage	Check and adjust line voltage to correct level
	3. Defective coil	Replace coil
	4. Wrong coil	Check manufacturer's recommendations; replace with correct coil
	5. Excessive magnet gap	Repair/replace worn components
Solenoid fails to drop out	1. Mechanical binding	Free linkage; repair/replace worn components
	2. Voltage still applied to coil	Check control to find reason for applied voltage
	3. Foreign material on solenoid	Clean solenoid; protect from contamination
Mechanical damage	1. Overvoltage to coil	Reduce voltage to proper level
	2. Insufficient spring pressure	Replace defective spring with correct size spring
	3. Abrasive material in relay	Clean relay; remove source of contamination
	4. Noisy relay (chattering)	See Noisy relay section

Thermally Operated Relays

Insulation failure	1. Overheating	Reduce current through relay; lower ambient temperature
	2. Contamination on relay	Clean relay; prevent additional contamination from accumulating on relay

Trouble	Possible Cause	Suggested Correction
	3. Moisture on relay	Allow to dry; shield from further moisture
Relay will not trip	1. Heater too large	Install correct size heater
	2. Mechanical bind of relay	Free mechanical bind; replace defective parts
	3. High ambient temperatures	Lower temperature; install high-temperature heater
	4. Contamination	Clean relay; shield from additional contamination
	5. Defective relay	Repair/replace relay
Relay will not reset	1. Defective relay	Repair/replace relay
	2. Worn components	Replace worn components
	3. Contamination	Clean relay; shield from additional contamination
Relay trips too low	1. Wrong size heater	Install correct heater
	2. High ambient temperature	Ventilate relay; install larger heater
Burning and welding of components	1. Excessive control current	Reduce current; use larger relays
	2. Contamination	Clean relay; shield from additional contamination
	3. Excessive external vibration	Reduce/remove source of vibration
	4. Misapplication	Use the relay for the designed application

Miscellaneous Circuit Components Including Resistors Capacitors, Fuses, Rectifiers, and Transformers

Trouble	Possible Cause	Suggested Correction
Insulation failures	1. Overheating	Lower the ambient temperature; reduce amount of current flow through the device

Trouble	Possible Cause	Suggested Correction
	2. Contamination	Clean/replace device; shield device from additional contamination
Overheating	1. Improper rated device	Check the device rating; use only for manufacturer's rated loads
Corrosion of device	1. Excess moisture	Dry device; shield from excess moisture
	2. Corrosive fumes	Repair/replace device; vent fumes away from device
Breakage or abnormal wear	1. Overheating	See Overheating section
	2. Physical abuse	Repair/replace device; guard against additional abuse
	3. External vibration	Isolate from vibration
	4. Electrical overloads	Reduce loads; protect from further overloads
Internal breakdown of device (primarily solid-state devices)	1. High temperatures	Reduce ambient temperature; reduce load through device
	2. Contamination (especially moisture or corrosive vapors)	Clean/replace device; protect the device from additional contaminants
	3. Improper handling and installation	Follow manufacturer's recommendations for handling and installation
Blowing too fast (fuses or circuit interrupters)	1. High ambient temperature	Lower ambient temperature; use higher-rated fuse (follow manufacturer's guidelines)
	2. Loose connections (wires, fuse clips, etc.)	Tighten all fittings; replace all damaged connectors
	3. Excessive current	Check circuit rating; reduce current to proper level
	4. Wrong fuse	Check manufacturer's recommendation for fuse size

Trouble	Possible Cause	Suggested Correction
Too slow interrupting circuit	1. Wrong sized device	Check manufacturer's recommendation for correct size of device

Switches, Including Limit, Float, Flow, Pressure, Push Button, and Selector

Trouble	Possible Cause	Suggested Correction
Occasional sticking of device	1. Contamination	Clean/replace device; shield from additional contaminants
	2. Worn components	Repair/replace worn components
	3. Improper adjustment	Properly adjust device for correct operation
	4. Mechanical binding	Check for free operation; position for free movement
Burning or welding of contact tips	1. High current across tips	Reduce current to proper level; replace tips with ones that have a high current rating
	2. Contamination	Replace tips; remove source of contamination; shield from further contamination
	3. External vibration	Isolate switch from vibration
Mechanical failures of switches	1. Contamination	Clean switch; protect from further contamination
	2. Wrong application for the switch	Consult manufacturer's specifications
	3. Handling, installation, or mechanical damage	Use proper installation and handling procedures; protect device from external mechanical damage
Failure to open or close contact tips	1. High current or voltage	Check circuit for correct values; use correct size device

Trouble	Possible Cause	Suggested Correction
	2. Inductive load in circuit	Consult manufacturer's recommendations for correct device
	3. Mechanical damage to switch	Repair/replace damaged components; protect switch from further damage
	4. Worn internal parts	Repair/replace parts
	5. Contamination	Clean switch; protect from further contamination
Leakage (flow and pressure switches)	1. Excessive pressure	Check switch rating; reduce system pressure
	2. Defective seals	Replace seals (or switch)
	3. Contamination	Clean switch; protect from further contamination
Electronic components	1. Overheating	Provide adequate ventilation and cooling. These devices will have a maximum life when ambient temperature is below 120°F. Failure rates increase approximately 100% for each 30°F increase of the ambient temperature

Appendix B Standard Elementary Diagram Symbols

Switches									
Disconnect	Circuit Interrupter	Circuit Breaker with Thermal OL	Circuit Breaker with Magnetic OL	Circuit Breaker with thermal and Magnetic OLs	Limit Switches		Foot Switches		
					Normally Open	Normally Closed	NO	NC	
					Held closed	Held open			

Pressure and Vacuum Switches		Liquid Level Switch		Temperature-Actuated Switch		Flow Switch (Air, Water, etc.)	
NO	NC	NO	NC	NO	NC	NO	NC

Speed (Plugging)		Anti-plug	Selector		
F	F	F	2 Position	3 Position	2 Pos. Sel. Push Button
R	R				

2 Position:

A1	X	
A2		X
	Low	High

o—o A1
o   o A2

3 Position:

A1	X		
A2			X
	Hand	Off	Auto

o   o A1
o   o A2

2 Pos. Sel. Push Button:

A1	X			
A2		X	X	X
	Free	Depres'd	Free	Depres'd
	Jog		Run	

o—o A1
o   o A2

Push Buttons						Plot Lights	
Momentary Contact				Maintained Contact		Indicate Color by Letter	
Single Circuit		Double Circuit		Mushroom Head	Two Single Circuits	One Double Circuits	Nonpush-to-Test	Push-to-Test
NO	NC	NO	NC					

Contacts				Coils		Overload Relays		Inductors
Instant Operating		Timed Contacts – Contact Action Retarded After Coil is:		Shunt	Series	Thermal	Magnetic	Iron Core
With Blowout		Without Blowout		Energized	Deenergized			
NO	NC	NO	NC	NOTC	NCTO	NOTO	NCTC	

Air Core

Transformers				ac Motors				dc Motors			
Auto	Iron Core	Air Core	Current	Dual Voltage	Single Phase	3-Phase Squirrel Cage	2 Phase 4 Wire	Wound Rotor	Armature	Shunt Field	Series Field	Commutator or Compensating Field

(Show 4 loops) | (Show 3 loops) | (Show 2 loops)

173

Appendix B   "Continued"

SPST, NO		SPST, NC		SPDT		Terms
Single Break	Double Break	Single Break	Double Break	Single Break	Double Break	SPDT = single pole, double throw
						SPST = single pole, single throw
DPST, 2 NO		**DPST, 2 NC**		**DPDT**		DPST = double pole, single throw
Single Break	Double Break	Single Break	Double Break	Single Break	Double Break	DPDT = double pole, double throw
						NO = normally open
						NC = normally closed

174

Appendix C Resistor Color Chart

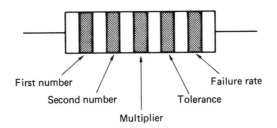

First number

Second number

Multiplier

Tolerance

Failure rate

Color	First Number	Second Number	Multiplier	Tolerance	Failure Rate per 1000 hr
Black	0	0	1	±20%	L 5%
Brown	1	1	10	± 1%	M 1%
Red	2	2	100	± 2%	P 0.1%
Orange	3	3	1,000	—	R 0.01%
Yellow	4	4	10,000	—	S 0.001%
Green	5	5	100,000	—	T 0.0001%
Blue	6	6	1,000,000	—	
Violet	7	7	10,000,000	—	
Gray	8	8	—	—	
White	9	9	—	—	
Gold	—	—	0.1	± 5%	
Silver	—	—	0.1	±10%	
No Color	—	—	—	±20%	

Appendix D American Wire Gauge (AWG)

Brown and Sharpe Gauge Number	Diameter (in mils)
#8	128
#9	114
#10	102
#11	91
#12	81
#13	72
#14	64
#15	57
#16	51
#17	45.3
#18	40.3
#19	35.9
#20	32.0
#21	28.5
#22	25.4
#23	22.6
#24	20.1

Index